Antibacterial Surfaces, Thin Films, and Nanostructured Coatings

Antibacterial Surfaces, Thin Films, and Nanostructured Coatings

Editor

Daniele Valerini

MDPI • Basel • Beijing • Wuhan • Barcelona • Belgrade • Manchester • Tokyo • Cluj • Tianjin

Editor
Daniele Valerini
Department for Sustainability
ENEA - Italian National Agency
for New Technologies, Energy
and Sustainable Economic
Development
Brindisi
Italy

Editorial Office
MDPI
St. Alban-Anlage 66
4052 Basel, Switzerland

This is a reprint of articles from the Special Issue published online in the open access journal *Coatings* (ISSN 2079-6412) (available at: www.mdpi.com/journal/coatings/special_issues/antibacterial_surf_film_nanostruct_coat).

For citation purposes, cite each article independently as indicated on the article page online and as indicated below:

LastName, A.A.; LastName, B.B.; LastName, C.C. Article Title. *Journal Name* **Year**, *Volume Number*, Page Range.

ISBN 978-3-0365-1632-5 (Hbk)
ISBN 978-3-0365-1631-8 (PDF)

Contents

Sebastian Balos, Tatjana Puskar, Michal Potran, Bojana Milekic, Daniela Djurovic Koprivica, Jovana Laban Terzija and Ivana Gusic
Modulus, Strength and Cytotoxicity of PMMA-Silica Nanocomposites
Reprinted from: *Coatings* **2020**, *10*, 583, doi:10.3390/coatings10060583 . . . . **127**

Luis Miguel Anaya-Esparza, Zuamí Villagrán-de la Mora, Noé Rodríguez-Barajas, Teresa Sandoval-Contreras, Karla Nuño, David A. López-de la Mora, Alejandro Pérez-Larios and Efigenia Montalvo-González
Protein–TiO_2: A Functional Hybrid Composite with Diversified Applications
Reprinted from: *Coatings* **2020**, *10*, 1194, doi:10.3390/coatings10121194 . . . . **141**

About the Editor

Daniele Valerini

Daniele Valerini, Ph.D., is a researcher at ENEA —Italian National Agency for New Technologies, Energy and Sustainable Economic Development, in Brindisi (Italy). He graduated in Physics in 2004 (full marks cum laude) and earned his Ph.D. in Physics in 2008. Since 2010 he is researcher with permanent position at ENEA, where his main R&D activities are related to development and characterization of coatings and nanostructured materials for antimicrobial applications, mechanical machining, transportation, energy and environment. His expertise is mainly focused on deposition of materials by PVD techniques and study of their morphological, structural, compositional, optical, electrical, mechanical, tribological and antimicrobial properties. Involved in several national and international projects, coordinator of the European network of infrastructure "EXTREME" and of the corresponding project funded by the EU KIC "EIT RawMaterials", and WG member of COST Action CA15102 "CRM-EXTREME". Organizer, committee member, chairman and speaker in several international events and symposia. Lecturer and trainer in R&D projects, professional training and University courses, and co-supervisor of bachelor students. Author of currently more than 60 international publications like peer-reviewed papers, reviews, editorials, conference proceedings, books and book chapters. Co-editor and guest editor of volume and special issue in international journals, and reviewer for more than 20 international journals.

Preface to "Antibacterial Surfaces, Thin Films, and Nanostructured Coatings"

Creating antibacterial surfaces is the primary approach in preventing the occurrence and diffusion of clinical infections and foodborne diseases as well as in contrasting the propagation of pandemics in everyday life. Proper surface engineering can inhibit microorganism spread and biofilm formation, can contrast antimicrobial resistance (AMR), and can avoid cross-contamination from a contaminated surface to another and eventually to humans. For these reasons, antibacterial surfaces play a key role in many applications, ranging from biomedicine to food and beverage materials, textiles, and objects with frequent human contact. The incorporation of antimicrobial agents within a surface or their addition onto a surface are very effective strategies to achieve this aim and to properly modify many other surface properties at the same time. In this framework, this Special Issue collects research studying several materials and methods related to the antibacterial properties of surfaces for different applications and discussions about the environmental and human-safety aspects.

As a guest editor of this Special Issue, I would like to express my gratitude to all of the contributing authors, to the reviewers, and to the supporting editors and hope that the papers included herein are of interest to readers and helpful in promoting further advances in the related R&D sectors.

Daniele Valerini
Editor

coatings

Editorial

Antibacterial Surfaces, Thin Films, and Nanostructured Coatings

Daniele Valerini

ENEA—Italian National Agency for New Technologies, Energy and Sustainable Economic Development, SSPT-PROMAS-MATAS, S.S. 7 Appia, km 706, 72100 Brindisi, Italy; daniele.valerini@enea.it

Citation: Valerini, D. Antibacterial Surfaces, Thin Films, and Nanostructured Coatings. *Coatings* **2021**, *11*, 556. https://doi.org/10.3390/coatings11050556

Received: 6 April 2021
Accepted: 4 May 2021
Published: 8 May 2021

Antibacterial surfaces can play a key role in a great number of everyday applications, spanning from biomedical purposes (medical devices, protection equipment, surgery tools, human implants, etc.) to usages for food and beverages (e.g., packaging). Such surfaces are fundamental to prevent the occurrence and diffusion of clinical infections and foodborne diseases, or to preserve the quality of the packaged content.

Different approaches can be pursued to confer antimicrobial properties to a given surface, such as the incorporation of antibacterial agents within the material surface or their deposition as coating films. Several organic (enzymes, natural extracts, etc.) and inorganic (metals, oxides, etc.) antibacterial agents, each with their own peculiar characteristics, are continuously studied and tested, seeking for ever-improved performance. In particular, in recent decades, new and ever more efficient materials have been experimented and effectively used in the aforementioned applications, especially after the enormous advances in nanotechnologies.

At the same time, together with the antimicrobial performance, for such kinds of applications, other essential aspects should be considered, such as the enhancement of other materials properties (mechanical, optical, wettability, etc.), as well as human safety issues, biocompatibility and environmental aspects.

In this framework, this Special Issue was aimed to collect original research works and reviews dealing with antibacterial surfaces and related aspects, from advances in materials and surface engineering, to characterization and functional properties, toxicity/safety for human health, and environmental aspects.

This Special Issue includes nine research papers and one review, relating to various strategies aimed at contrasting bacterial proliferation on surfaces or studying the biocompatibility of materials for many different applications.

In the first research work [1], different amounts of silica (SiO_2) nanoparticles were added into Poly(methyl methacrylate) (PMMA) to fabricate a reinforced nanocomposite material with improved mechanical properties, evaluating its cytotoxic effects with the aim to get a biocompatible material with possible applications in dentures.

A material widely employed in implants is titanium, where it needs to exhibit biocompatibility and antimicrobial properties. In work [2], titanium foils were coated with a blend made of chitosan (CS) and zinc-coated halloysite nanotubes (ZnHNTs) loaded with gentamicin sulfate. Such composite coating was tested in order to get a biocompatible material able to provide antimicrobial action against *Staphylococcus aureus* thanks to the release of metal ions and antibiotic gentamicin.

Similarly, another composite material, which could deliver possible controlled release of antimicrobial agents, was studied in the work presented in paper [3]. Here, a photo-electrochemical reduction process was employed to synthesize silver nanoparticles (Ag NPs) in a hydroxypropyl methylcellulose (HPMC) matrix, whose bactericidal effects were tested against *Escherichia coli* and *Staphylococcus epidermidis* species. This system was then expected to represent a promising material for the controlled release of antibacterial Ag NPs through the slow dissolution of HPMC in water.

With the aim to prevent formation of biofilms of multi-drug resistant pathogens, the authors of paper [4] developed zinc oxide nanoparticles (ZnO NPs) functionalized with extracellular polysaccharide xanthan gum (XG) by a green route. The quorum sensing inhibitory activity of this nanocomposite was evaluated against Gram-negative pathogens *Chromobacterium violaceum* and *Serratia marcescens*, demonstrating a significant inhibition of biofilm formation.

With the same aim to develop new materials against antimicrobial-resistant (AMR) bacteria, in the work presented in paper [5] different metal ions (Ag, Au, Pd, Pt, Zn, Ga) were tested alone and in combination with graphene matrices acting as metal ion carriers. The antibacterial action of these systems was evaluated against antibiotic susceptible and antibiotic resistant strains of *Acinetobacter baumannii*, *Staphylococcus aureus*, *Klebsiella pneumoniae* and *Pseudomonas aeruginosa*, demonstrating the greatest activity with Au, Pd and Pt.

In order to develop eco-friendly materials with wide-spectrum antimicrobial activity for possible use in biomedical and food packaging applications, the work presented in [6] investigated the combination of two antibacterial agents—aluminum doped zinc oxide (Al-doped ZnO, abbr. AZO) and silver (Ag)—in nanostructured layers deposited by different methods onto bioplastic polylactide (PLA) films. The two active agents showed preferential antibacterial activity against Gram-positive *Staphylococcus aureus* and Gram-negative *Escherichia coli* species, respectively, so that their synergistic dual action in the combined coatings was able to provide a total bacterial suppression against both species.

Coatings based on naturally occurring antimicrobial peptides (AMPs) were tested in [7] to produce antibacterial surfaces for biomaterials. Different AMPs (LL37, Magainin 2, and Parasin 1) were covalently grafted onto an aldehyde plasma polymer (ALDpp), acting as a polymer interlayer that can be deposited onto different surfaces. Considerable reduction in bacterial colonization was demonstrated with *Staphylococcus epidermidis*, *Staphylococcus aureus* and *Escherichia coli*, while no significant cytotoxicity was found to primary human fibroblasts.

Another chemical-based route was employed in [8] to synthesize PLA-based amphiphilic copolymer micelles loaded with a photosensitizer (PS) of the BODIPY dyes family. Such micelles were sprayed on glass substrates to form a coating able to release BODIPY, resulting in an antibacterial action against *Staphylococcus aureus*.

In another work [9], the combination of the electrospinning method and the physical sputtering deposition technique was successfully used to produce antibacterial functionalized fiber scaffolds. In particular, Ag nanoparticles were sputter-deposited on electrospun polycaprolactone (PCL) fiber mats, without any significant damage induced on the soft polymer fibers, allowing the conferment of excellent antibacterial activity against *Escherichia coli*. The easy and flexible fabrication of these PCL-Ag mats can be applicable to several sectors, such as biomedical devices, bioremediation and antifouling systems in filtration, personal protective equipment (PPE), food packaging, etc.

Finally, a review [10] was presented about the use of hybrid organic–inorganic composites made by the incorporation of titanium dioxide (TiO_2) nanoparticles in protein-based materials. The characteristics of such compounds, including their antimicrobial properties, were discussed for possible usages in various applications (packaging, biomedical, pharmaceutical, environmental remediation, textiles, etc.), together with considerations on the issues related to TiO_2 concentrations and need for standardization in production protocols.

As shown by the various works published in this Special Issue, it is clear how research efforts are continuously conducted by means of different strategies to provide antibacterial properties to surfaces for various applications. Many different materials are proposed, including organic and inorganic ones, as well as hybrid blends. Additionally, many different fabrication techniques can be employed, such as physical-based or chemical-based routes and their combination. At the same time, many aspects would require further examination, for example, to deeply assess the safety to human beings, the stability of the materials, and the duration of the antimicrobial activity.

Acknowledgments: D.V. would like to thank all the Authors for their valuable contributions to this Special Issue, the Reviewers for their reviews and useful comments allowing the improvement of the submitted papers, and the Journal Editors for their kind support throughout the production of this Special Issue.

Conflicts of Interest: The author declares no conflict of interest.

References

1. Balos, S.; Puskar, T.; Potran, M.; Milekic, B.; Djurovic Koprivica, D.; Laban Terzija, J.; Gusic, I. Modulus, Strength and Cytotoxicity of PMMA-Silica Nanocomposites. *Coatings* **2020**, *10*, 583. [CrossRef]
2. Humayun, A.; Luo, Y.; Mills, D.K. Electrophoretic Deposition of Gentamicin-Loaded ZnHNTs-Chitosan on Titanium. *Coatings* **2020**, *10*, 944. [CrossRef]
3. Kvitek, O.; Mutylo, E.; Vokata, B.; Ulbrich, P.; Fajstavr, D.; Reznickova, A.; Svorcik, V. Photochemical Preparation of Silver Colloids in Hydroxypropyl Methylcellulose for Antibacterial Materials with Controlled Release of Silver. *Coatings* **2020**, *10*, 1046. [CrossRef]
4. Husain, F.M.; Hasan, I.; Qais, F.A.; Khan, R.A.; Alam, P.; Alsalme, A. Fabrication of Zinc Oxide-Xanthan Gum Nanocomposite via Green Route: Attenuation of Quorum Sensing Regulated Virulence Functions and Mitigation of Biofilm in Gram-Negative Bacterial Pathogens. *Coatings* **2020**, *10*, 1190. [CrossRef]
5. Slate, A.J.; Karaky, N.; Crowther, G.S.; Butler, J.A.; Banks, C.E.; McBain, A.J.; Whitehead, K.A. Graphene Matrices as Carriers for Metal Ions against Antibiotic Susceptible and Resistant Bacterial Pathogens. *Coatings* **2021**, *11*, 352. [CrossRef]
6. Valerini, D.; Tammaro, L.; Vigliotta, G.; Picariello, E.; Banfi, F.; Cavaliere, E.; Ciambriello, L.; Gavioli, L. Ag Functionalization of Al-Doped ZnO Nanostructured Coatings on PLA Substrate for Antibacterial Applications. *Coatings* **2020**, *10*, 1238. [CrossRef]
7. Griesser, S.S.; Jasieniak, M.; Vasilev, K.; Griesser, H.J. Antimicrobial Peptides Grafted onto a Plasma Polymer Interlayer Platform: Performance upon Extended Bacterial Challenge. *Coatings* **2021**, *11*, 68. [CrossRef]
8. Caruso, E.; Orlandi, V.T.; Malacarne, M.C.; Martegani, E.; Scanferla, C.; Pappalardo, D.; Vigliotta, G.; Izzo, L. Bodipy-Loaded Micelles Based on Polylactide as Surface Coating for Photodynamic Control of *Staphylococcus aureus*. *Coatings* **2021**, *11*, 223. [CrossRef]
9. Valerini, D.; Tammaro, L.; Vitali, R.; Guillot, G.; Rinaldi, A. Sputter-Deposited Ag Nanoparticles on Electrospun PCL Scaffolds: Morphology, Wettability and Antibacterial Activity. *Coatings* **2021**, *11*, 345. [CrossRef]
10. Anaya-Esparza, L.M.; Villagrán-de la Mora, Z.; Rodríguez-Barajas, N.; Sandoval-Contreras, T.; Nuño, K.; López-de la Mora, D.A.; Pérez-Larios, A.; Montalvo-González, E. Protein–TiO_2: A Functional Hybrid Composite with Diversified Applications. *Coatings* **2020**, *10*, 1194. [CrossRef]

coatings

MDPI

Article

Graphene Matrices as Carriers for Metal Ions against Antibiotic Susceptible and Resistant Bacterial Pathogens

Anthony J. Slate [1,2], Nathalie Karaky [2], Grace S. Crowther [3], Jonathan A. Butler [3], Craig E. Banks [3], Andrew J. McBain [4] and Kathryn A. Whitehead [2,*]

[1] Department of Biology and Biochemistry, University of Bath, Claverton Down, Bath BA2 7AY, UK; ajs319@bath.ac.uk
[2] Microbiology at Interfaces, Manchester Metropolitan University, Chester Street, Manchester M1 5GD, UK; NATHALIE.KARAKY@stu.mmu.ac.uk
[3] Faculty of Science and Engineering, Manchester Metropolitan University, Chester Street, Manchester M1 5GD, UK; gracescrowther@gmail.com (G.S.C.); Jonathan.Butler@mmu.ac.uk (J.A.B.); c.banks@mmu.ac.uk (C.E.B.)
[4] Division of Pharmacy and Optometry, Faculty of Biology, Medicine and Health, School of Health Sciences, The University of Manchester, Manchester M13 9PL, UK; andrew.mcbain@manchester.ac.uk
* Correspondence: k.a.whitehead@mmu.ac.uk; Tel.: +44-(0)161-247-1157

Abstract: Due to the ever-increasing burden of antimicrobial-resistant (AMR) bacteria, the development of novel antimicrobial agents and biomaterials to act as carriers and/or potentiate antimicrobial activity is essential. This study assessed the antimicrobial efficacy of the following ionic metals, silver, gold, palladium, platinum, zinc, and gallium alone and in combination with graphene matrices (which were coated via a drop casting coating method). The graphene foam was utilized as a carrier for the ionic metals against both, antibiotic susceptible and resistant bacterial strains of *Acinetobacter baumannii, Staphylococcus aureus, Klebsiella pneumoniae* and *Pseudomonas aeruginosa*. Ionic gold, palladium and platinum demonstrated the greatest antimicrobial activity against the susceptible and resistant strains. Scanning electron microscopy (SEM) visualized cellular ultrastructure damage, when the bacteria were incubated upon the graphene foam alone. This study suggests that specific metal ions applied in combination with graphene foam could present a potential therapeutic option to treat AMR bacterial infections. The application of the graphene foam as a potential carrier could promote antimicrobial activity, provide a sustained release approach and reduce possible resistance acquisition. In light of this study, the graphene foam and ionic metal combinations could potentially be further developed as part of a wound dressing.

Keywords: metal ions; graphene; antibiotic resistance; foams; biomaterials

Citation: Slate, A.J.; Karaky, N.; Crowther, G.S.; Butler, J.A.; Banks, C.E.; McBain, A.J.; Whitehead, K.A. Graphene Matrices as Carriers for Metal Ions against Antibiotic Susceptible and Resistant Bacterial Pathogens. *Coatings* **2021**, *11*, 352. https://doi.org/10.3390/coatings11030352

Academic Editor: Daniele Valerini

Received: 29 January 2021
Accepted: 16 March 2021
Published: 19 March 2021

1. Introduction

Healthcare-associated infections (HAIs) are a substantial burden on healthcare settings worldwide. The emergence of multidrug resistant (MDR) bacterial strains has further exacerbated this problem, resulting in serious financial burdens on healthcare services [1–3]. The lack of effective antimicrobial therapies to combat MDR bacterial strains has led to an urgent need to develop alternative therapeutic options [4].

It is estimated that in developed countries chronic wounds occur in around 2 % of the population [5]. Delayed wound healing can lead to both local and systemic complications such as bacteraemia, osteomyelitis, sepsis and ultimately death [6]. Wound dressings play a pivotal role in the management of wound healing. The use of antimicrobial agents within the dressing structure, such as silver sulfadiazine, has been previously associated with a positive outcome, resulting in a reduction in bacterial infections [7]. Other wound dressings such as those which incorporate silver nanoparticles can also promote cellular proliferation and therefore wound healing, whilst maintaining high levels of antimicrobial efficacy [8,9].

Modern examples of metals utilised as therapeutic agents include cisplatin-based anti-cancer drugs, anti-arthritis drugs and topical antimicrobials (such as zinc salts) [10].

Different metal-based compounds have distinct structural configurations (e.g., ions, oxides and nanoparticles) which exhibit different mechanisms of antimicrobial action. The most common mechanisms of action however include, cellular ultrastructure damage via oxidative stress, membrane damage and protein dysfunction. Importantly, the application of metals as antimicrobial agents may result in a reduced probability of resistance occurrence; in contrast to many antibiotics, metals target multiple components of the bacterial cell, simultaneously [11–13].

Graphene is defined as a two-dimensional (2D) monolayer lattice of sp^2 hybridised carbon atoms [14–16]. Graphene has unique properties such as excellent thermal and electrical conductivity, permeability to gasses, excellent tensile strength and it can be readily chemically functionalized, such properties have resulted in a plethora of applications [17]. Since its discovery in 2004, graphene and its derivatives, have been the focus of numerous research groups for a myriad of applications including electrochemistry (e.g., electrodes, sensors and supercapacitors), drug delivery (graphene-based drug carriers), dental fillers, water/surface disinfection and antimicrobial activity [18–23].

Antimicrobial properties have been demonstrated by graphene and its derivatives [24]. Graphene is believed to perturb the bacterial cytoplasmic membrane, due to the insertion of sharp graphene edges, leading to the loss of membrane integrity, resulting in cell lysis and death [25]. Other proposed antimicrobial mechanisms of graphene include wrapping, oxidative stress (with and without reactive oxygen species (ROS) production), lipid bilayer extraction and interference of protein-protein interactions [19,22]. Wrapping by graphene occurs when the microorganism is surrounded by graphene and nutritional and physicochemical conditions are altered [26]. Previously, simulations have demonstrated that graphene sheets > 5.20 nm could partially wrap around a bacterial species, resulting in the inversion of the phospholipid bilayer whilst, graphene sheets < 5.20 nm could penetrate the cell membrane [27]. A recent study conducted in 2020 by Butler et al., revealed that graphene derivatives (graphite, graphene and graphene oxide) potentiated the activity of antibiotics (ciprofloxacin, chloramphenicol and piperacillin/tazobactam) against *Enterococcus faecium, Klebsiella pneumoniae and Escherichia coli* [28].

Due to the increasing prevalence of MDR pathogens and scarcity of novel effective antibiotics, there has been very few major classes of broad-spectrum antibiotics developed over the last 40 years [29]. However, teixobactin, which was first discovered in 2015 is an effective antimicrobial agent against *S. aureus* and *Mycobacterium tuberculosis*; this antibiotic inhibits cell wall synthesis by binding to a highly conserved motif of lipid II (a precursor of peptidoglycan) and lipid III (a precursor of cell wall teichoic acid) [30]. Whilst, recent research conducted by Picconi et al., (2020) may have discovered a new broad-spectrum antibiotic class in the form of modified pyrrolobenzodiazepines with a C8-linked aliphatic heterocycle [31]. This novel broad-spectrum antibiotic class demonstrated potent antimicrobial activity against MDR Gram-negative bacteria [31]. Whilst the emergence of new broad-spectrum antibiotics is promising, alternative therapies antimicrobial therapies must also be considered in order to effectively treat MDR infections. The utilisation of metals and graphene-based compounds may be one potential avenue to explore in order to alleviate the burden placed on traditional therapeutic options. Furthermore, the use of metal ions in combination with graphene (in this instance graphene coated with ionic metals to act as a carrier) could potentially promote antimicrobial activity, therefore acting as an alternative wound dressing treatment.

This study aimed to investigate the antimicrobial activity of metal ions in combination with graphene foams via a drop cast coating method in order to determine the antimicrobial efficacy, whilst demonstrating the efficacy of the graphene foams as therapeutic carriers against a range of antibiotic susceptible and resistant bacteria.

2. Materials and Methods

2.1. Bacterial Strains

Four susceptible bacterial isolates, *Staphylococcus aureus* strain NCTC 12973, *Klebsiella pneumoniae* strain NCTC 9633, *Acinetobacter baumannii* strain NCTC 12156 and *Pseudomonas aeruginosa* strain NCTC 10332 and four resistant isolates, *S. aureus* hospital isolate 252, *K. pneumoniae* hospital isolate 1411061, *A. baumannii* isolate A483 and *Pseudomonas aeruginosa* strain VTK106689 were originally sourced from Leeds General Infirmary, UK. All strains were cultured on Tryptone Soya agar (TSA) or broth (TSB) and incubated for 24 h at 37 °C unless stated otherwise. Before experimentation all strains were characterised for antibiotic resistance using both Gram-positive and Gram-negative/urine MASTRING-S® (Gram-positive: M13/NCE, Gram-negative/Uropathogens M26/NCE, Mast Group, Merseyside, UK). Briefly, 100 µL of bacterial suspension (optical density (OD) 1.0 (±0.1) at 540 nm) was spread onto TSA and incubated at 37 °C for 18 h. The normalisation of the OD 540 nm resulted in bacterial concentrations of ca. 5.0×10^8 CFU mL^{-1}. Negative controls (containing no bacteria) were included to confirm sterility (data not shown). Following incubation, zones of inhibition (ZoIs) were measured using digital calipers accurate to three decimal places, in order to determine antibiotic resistance profiles. Two bacterial isolates were included for each bacterial species, the bacterial strain that demonstrated the greatest ZoI for all relevant antibiotics (e.g., *S. aureus* and common Gram-positive antibiotics) was catergorised as susceptible, whilst the other was deemed resistant.

2.2. Graphene Foam Preparation

Three-dimensional multilayer graphene foam was procured from Graphene-Supermar ket, USA and sections were prepared using a sterile cork-borer (7 mm diameter) and then sterilised for 2 h under UV light at 375 nm prior to experimentation. A drop cast method was used to incorporate the metal ions into the graphene foam. Aliquots of 20 µL of the test metal ion/acid (as described in the following section, "test compounds") were added directly to the surface of the graphene foam and left at room temperature for 30 min prior to experimentation.

2.3. Test Compounds

The metal ions evaluated in this study, silver, gold, palladium, platinum, zinc and gallium, were Atomic Absorption Standards (AAS) at a concentration of ca. 1000 µg mL^{-1} (1000 ppm) and were used without further modification (Merck, Feltham, UK). The solvents used to solubilise the metal ions were tested individually throughout this study and results were subtracted, in order to assure the antimicrobial effects observed were from the metal ions alone. Graphene foams (Graphene Supermarket, Reading, MA, USA) were also evaluated individually and in combination with the aforementioned metals.

2.4. Disc Diffusion Assays

The metal ions were evaluated individually and in combination with graphene foam (*via* a drop cast coating method) to determine their antimicrobial efficacy against the bacterial isolates. Bacterial strains were inoculated into TSB and incubated for 24 h at 37 °C with agitation (150 rpm). Overnight cultures were then adjusted to an OD of 1.0 (±0.1) at 540 nm. One hundred microliters of the adjusted suspension were inoculated onto a sterile TSA plate and spread to establish confluency. The graphene foams (pre-coated with metal ions) were then transferred to the centre of the inoculated agar plate. The antimicrobial activity of the metal solutions/acid controls in the absence of graphene was also evaluated by adding 20 µL of each of the metal ion solutions onto a sterile filter paper (6 mm diameter), this was placed into the centre of the inoculated agar. All plates were incubated at 37 °C for 24 h and zone of inhibition (ZoI) was established by measuring the inhibition diameter, using digital callipers accurate to 0.001 mm. This experiment was performed in triplicate from three independent starting cultures. Note, that the ZoI of the metal solution was subtracted from the ZoI of its corresponding metal ion to account for any antimicrobial effect of the acid.

2.5. Scanning Electron Microscopy

Scanning electron microscopy (SEM) was used to visualise the bacterial cells post-incubation with the graphene foam to determine if damage to the cellular ultrastructure was a potential mechanism of action. Following incubation upon silicon wafers or graphene foam (24 h at 37 °C), samples were dried at room temperature in a Class 2 biosafety cabinet (Atlas Clean Air, Preston, UK). The samples were prepared as per Butler et al. (2020) [32]. Briefly, the samples were fixed in 4% v/v glutaraldehyde for 24 h at 4 °C. Once fixed the samples were rinsed with sterile deionized water and subjected to an ethanol gradient, 10%, 30%, 50%, 70%, 90% and 100% v/v absolute ethanol. The samples were then stored in a desiccator over 24 h to remove any moisture and were the sputter coated with gold for 30 s (Polaron, London, UK) using the following parameters (power 5 mA, 30 s, 800 V, vacuum 0.09 mbar, argon gas). Scanning electron microscopy (SEM) was then performed using a JEOL (Tokyo, Japan) JSM-5600LV model SEM.

2.6. Statistical Analysis

Graphs were generated using Prism (Graphpad Software; version 8.4.3). The standard error of the mean (SEM) was denoted via error bars. For statistical analysis, p values were calculated at the 95 % confidence level by two-way ANOVAs, this was determined using Graphpad Prism (version 8). In all cases, $p < 0.05$ was considered statistically significant. Asterisks denote significance, * $p \leq 0.05$, ** $p \leq 0.01$, *** $p \leq 0.001$ and **** $p \leq 0.0001$.

3. Results

3.1. Antibiotic Susceptibility Profiles

The antibiotic susceptibility of the susceptible and resistant bacterial strains was determined against common broad-spectrum antibiotics that demonstrate antimicrobial activity against both Gram-positive (Figure 1A,B) and Gram-negative (Figure 2A,B) bacteria. When the antibiotics commonly utilised against Gram-positive bacteria were tested, *S. aureus* was the most susceptible strain, with eight out of eight antibiotics inhibiting the growth of this strain (NCTC 12973), whilst five out of the eight antibiotics (chloramphenicol, fusidic acid, novobiocin, streptomycin and tetracycline) resulted in growth inhibition of the resistant strain (Strain 252). The greatest inhibition of growth observed was by 10 µg fusidic acid against the resistant strain (Strain 252), producing an average ZoI of 32.57 mm. The least affected bacterial species to the commonly utilised antibiotics against Gram-positive bacteria was *A. baumannii* with only tetracycline producing a ZoI (16.90 mm) against the susceptible strain (NCTC 12156).

The MASTRING-S® antibiotics (M26/NCE) which are commonly utilised against urinary tract pathogens/Gram-negative bacteria were tested against the susceptible and resistant bacterial strains (Figure 2A,B). The growth of the susceptible strains, *S. aureus* (NCTC 12973) and *K. pneumoniae* (NCTC 9633) was the most affected with both exhibiting growth inhibition when tested against seven of the eight antibiotics. The largest ZoI (26.80 mm) was determined by 10 µg tetracycline against the susceptible *S. aureus* strain. The resistant bacterial strains demonstrated greater resistance profiles; five out of the eight antibiotics inhibited *S. aureus* (Strain 252) growth, whilst three of the eight inhibited *K. pneumoniae* (Strain 1411061). The antibiotic which demonstrated the greatest growth inhibition observed was 10 µg tetracycline (31.30 mm) produced against the resistant *S. aureus* (Strain 252). Overall, the most resistant strains utilised throughout this study were the resistant *A. baumannii* (LMDR A483) strain, with only two of the fourteen antibiotics inhibiting growth, and *P. aeruginosa* (VTK 106689) with only three antibiotics producing a ZoI. European Committee on Antimicrobial Susceptibility Testing (EUCAST) guidelines [33] were consulted to determine antimicrobial susceptibility (Electronic Supplementary Information, Table S1).

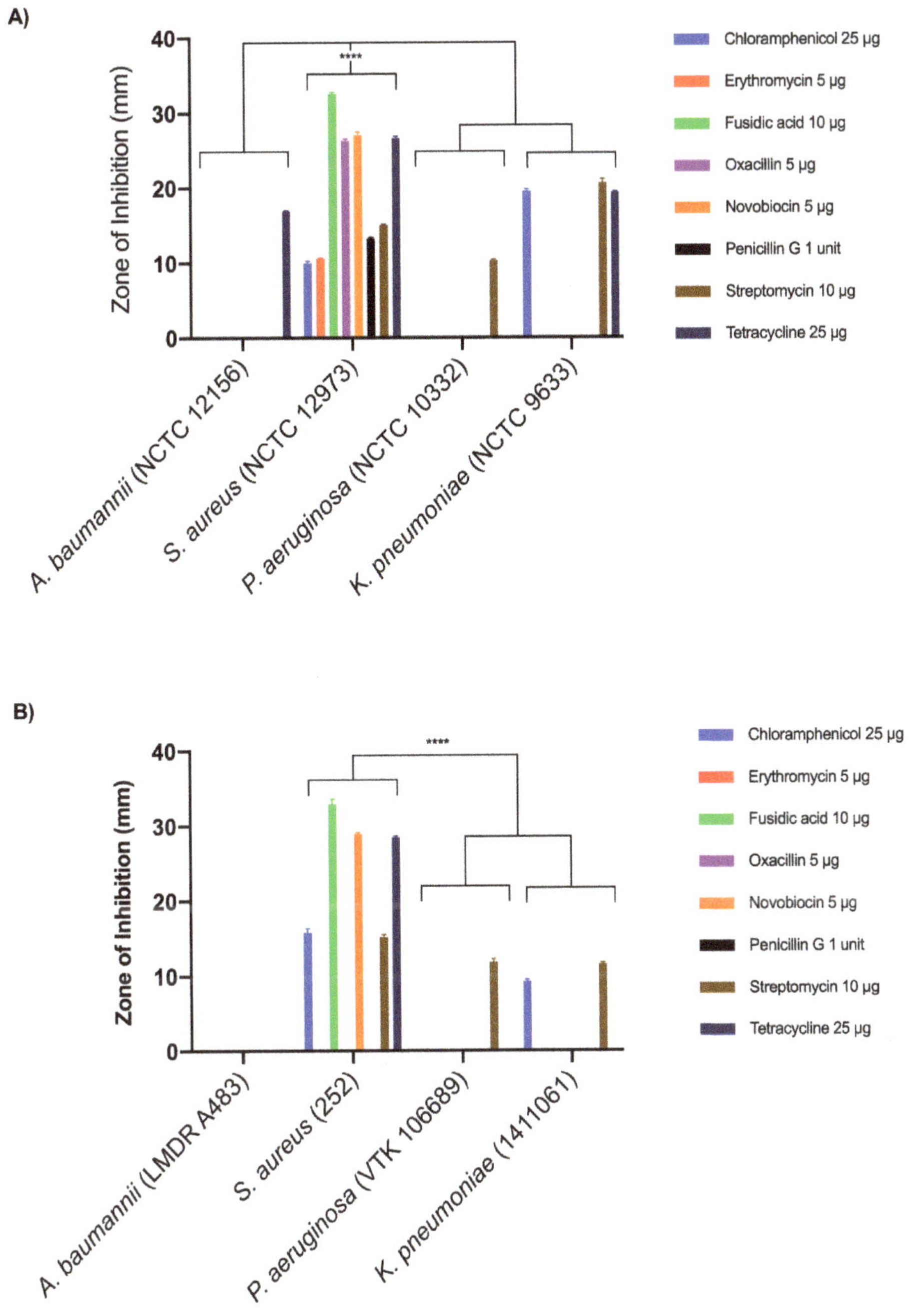

Figure 1. Antimicrobial susceptibility profiles against commonly used antibiotics against Gram-positive bacteria determined using MASTRING-S® (M13/NCE) against, (**A**) susceptible bacterial strains and (**B**) resistant bacterial strains (n = 3) **** $p \leq 0.0001$.

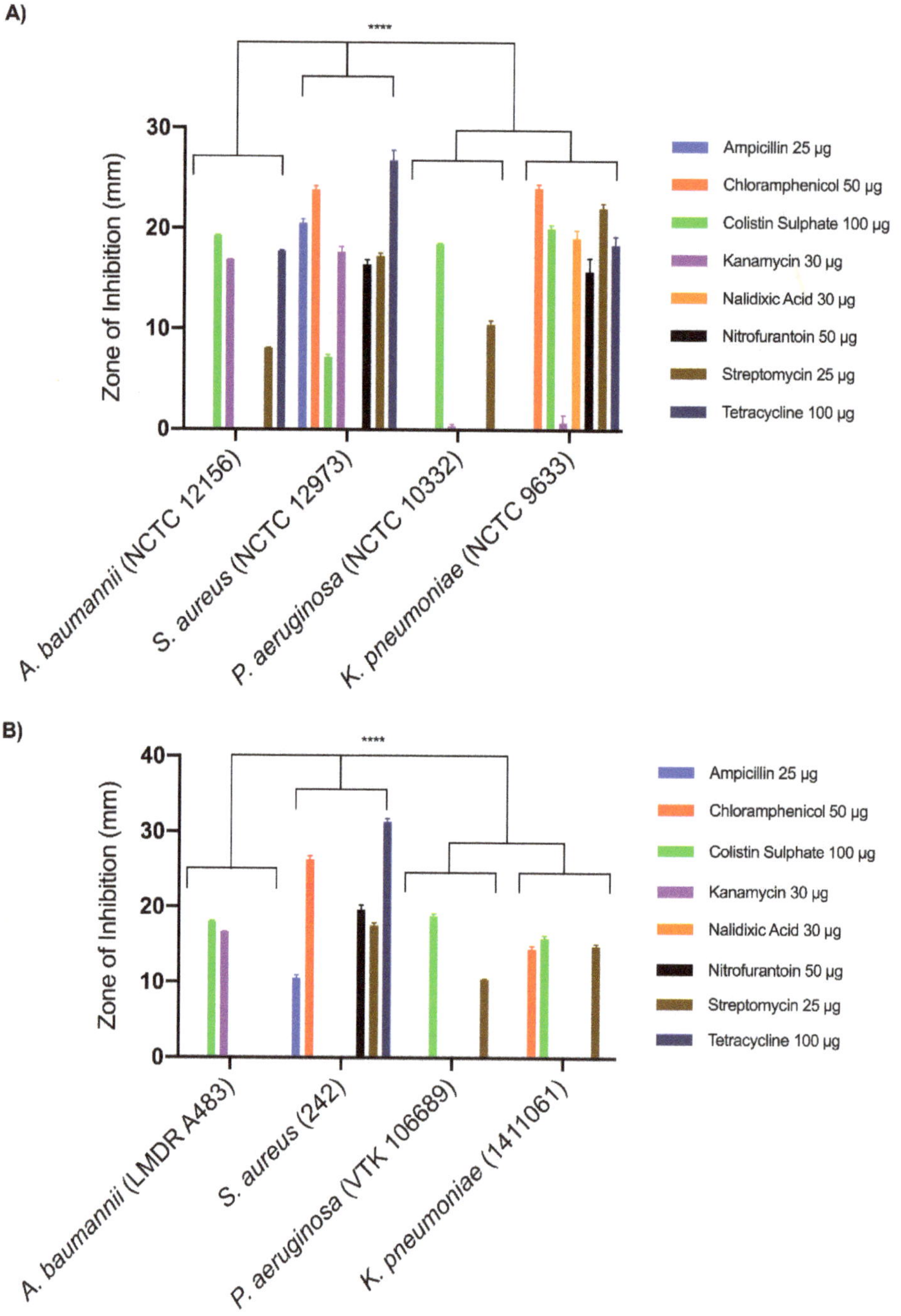

Figure 2. Antimicrobial susceptibility profiles against commonly used antibiotics against Gram-negative bacteria was determined using MASTRING-S® (M26/NCE) against, (**A**) susceptible bacterial strains and (**B**) resistant bacterial strains (n = 3) **** $p \leq 0.0001$.

3.2. Disc Diffusion Assays

Metal Ions Alone and Graphene Alone

Overall, palladium and gold displayed the greatest antimicrobial efficacy against the eight susceptible and resistant isolates tested (Figure 3). Both metals were able to inhibit the growth of most of the isolates producing ZoIs in the range of 14.10–19.53 mm and 15.14–17.68 mm against the susceptible isolates, and 12.73 mm–14.21 mm and 12.17–13.877 mm against the resistant strains, with palladium and gold, respectively. Silver, zinc and platinum similarly showed, good antimicrobial activity against the majority of the isolates (Figure 3A,B). The degree of antimicrobial activity by the metals tested observed was greater against the susceptible bacterial strains when compared to the resistant strains. Gallium exhibited the least antimicrobial activity of the metals tested against the bacterial isolates tested (Figure 3A,B). Interestingly, on average the metals produced greater inhibition of growth against the susceptible bacterial strains. Graphene exhibited the least antimicrobial activity producing ZoIs in the range of 7.00–7.78 mm for the susceptible strains and 7.82 mm–8.36 mm for the resistant strains.

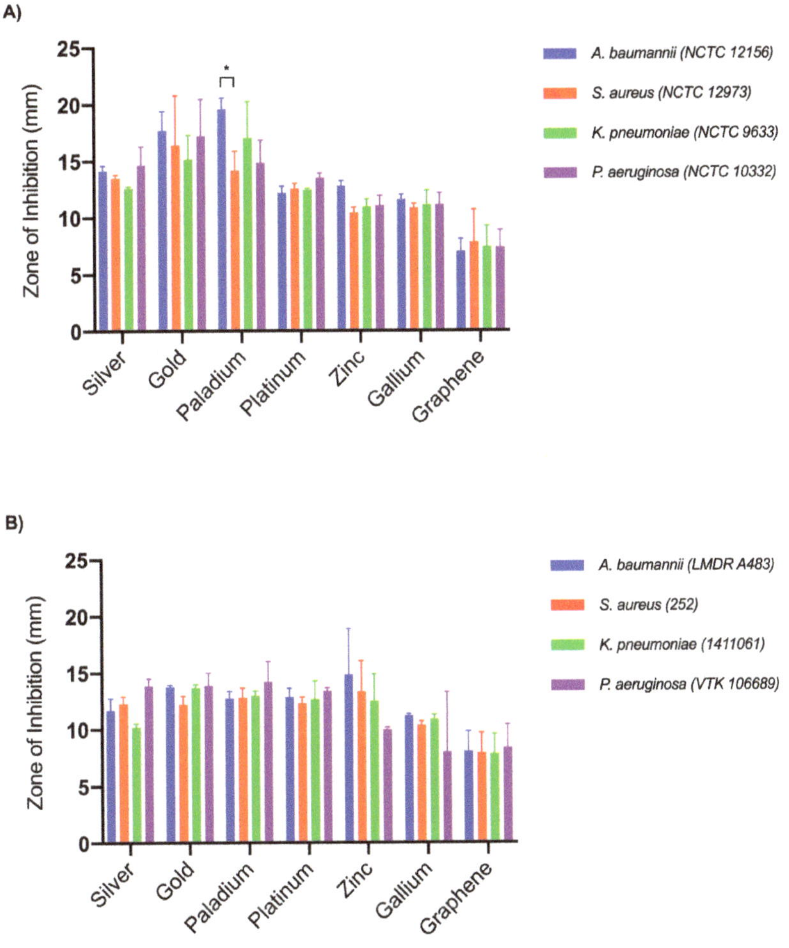

Figure 3. Zone of Inhibition determined by the metal ions and graphene alone against a range of (**A**) susceptible and (**B**) resistant bacteria (n = 3) * $p \leq 0.05$.

3.3. Metal Ions and Graphene in combination with Graphene Foam

To assess potentially synergistic activity of the metal ions when incorporated with graphene matrixes, zone of inhibition assays were conducted (Figure 4). In these assays, the graphene foam was used as a carrier for the metal ions. The antimicrobial efficacy demonstrated was similar to the metal ions tested alone, this indicated that the graphene did not have an antagonistic effect (Figure 3). Against the susceptible bacterial strains ionic silver in the presence of graphene demonstrated increased activity against *S. aureus* (14.60 mm; Figure 4A). Whilst, ionic platinum demonstrated slight synergistic activity with the graphene foams against the susceptible strains, producing enhanced ZoIs against *A. baumannii* (14.17 mm), *K. pneumoniae* (13.88 mm) and *P. aeruginosa* (14.46 mm) (Figure 4A).

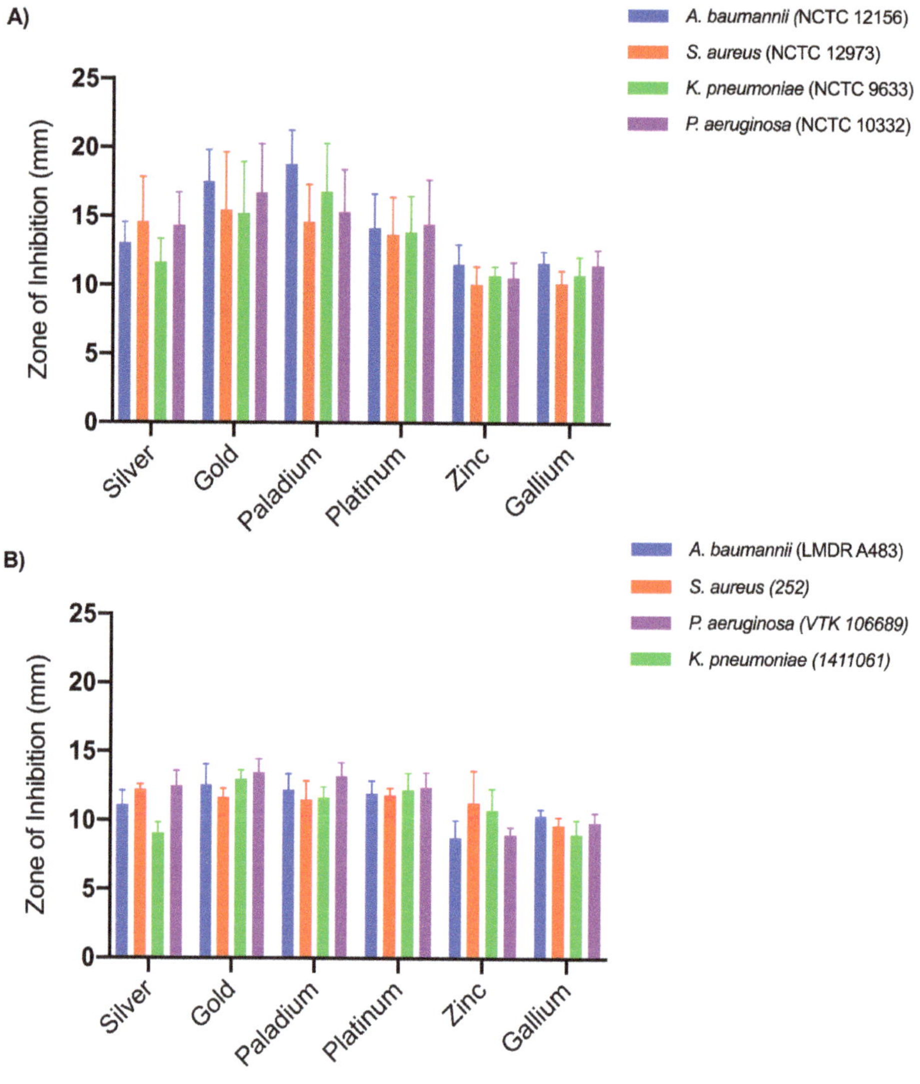

Figure 4. Zones of inhibition of graphene foams with metals on (**A**) antibiotic susceptible and (**B**) resistant bacteria (n = 3).

The ZoI assays combining the metal ions with the graphene foam was then conducted with the resistant bacterial strains (Figure 4B). Overall, the addition of the metal ions to the graphene foams resulted in similar antimicrobial activity as observed by the metal ions alone, with no detriment to antimicrobial efficacy observed, against both the susceptible and resistant bacterial strains.

3.4. Scanning Electron Microscopy

To visualise the effect of the graphene foams on the cellular ultrastructure of the bacterial strains, SEM was conducted. Firstly, the structure of the graphene foam was visualised at a low (×80) and higher magnification (×5000) (Figure 5). At a lower magnification, a rough surface topography was observed with pits which were larger than single microbial cell dimensions, this indicates that the bacteria could become entrapped in the graphene structure.

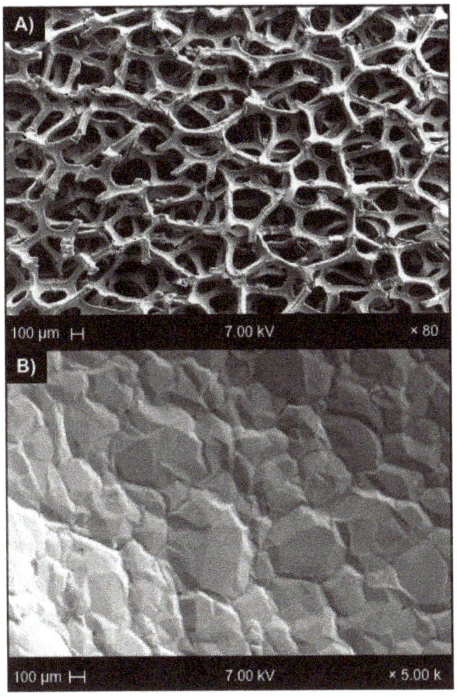

Figure 5. Scanning electron microscopy (SEM) images of (**A**) low (×80) and (**B**) high (×5000) magnification of graphene foam.

The bacterial strains were incubated for 24 h (at 37 °C) on both silicon wafers and graphene foams, the morphology of the resistant bacterial cells was then visualised (Figure 6). The cells incubated on the silicon wafers were used as the control to effectively compare against *S. aureus*, *K. pneumoniae*, *A. baumannii* and *P. aeruginosa* (Figure 6A, 6B, 6C and 6D, respectively) incubated on the graphene foams. Following incubation on the graphene foam (Figure 6E), *S. aureus* was compared against the bacteria incubated on silicon wafers (Figure 6A), cellular debris was observed in the presence of the graphene foams (Figure 6F and 6G), whilst, some bacterial cells had developed pertusions through the cellular membrane (Figure 6E), which were not evident in the control bacterial sample. With the Gram-negative bacterial species, there was less of a pronounced effect observed, except for with the *K. pneumoniae* and *A. baumannii* cells (Figure 6F,G). When incubated on the graphene foams the *A. baumannii* cells are much smaller than the control which could indicate cellular leakage. The *K. pneumoniae* cells demonstrated a similar morphology to the *A. baumannii* cells when incubated on graphene foams (Figure 6F). The *P. aeruginosa* cells were largely unaffected by incubation with the graphene foams (Figure 6H).

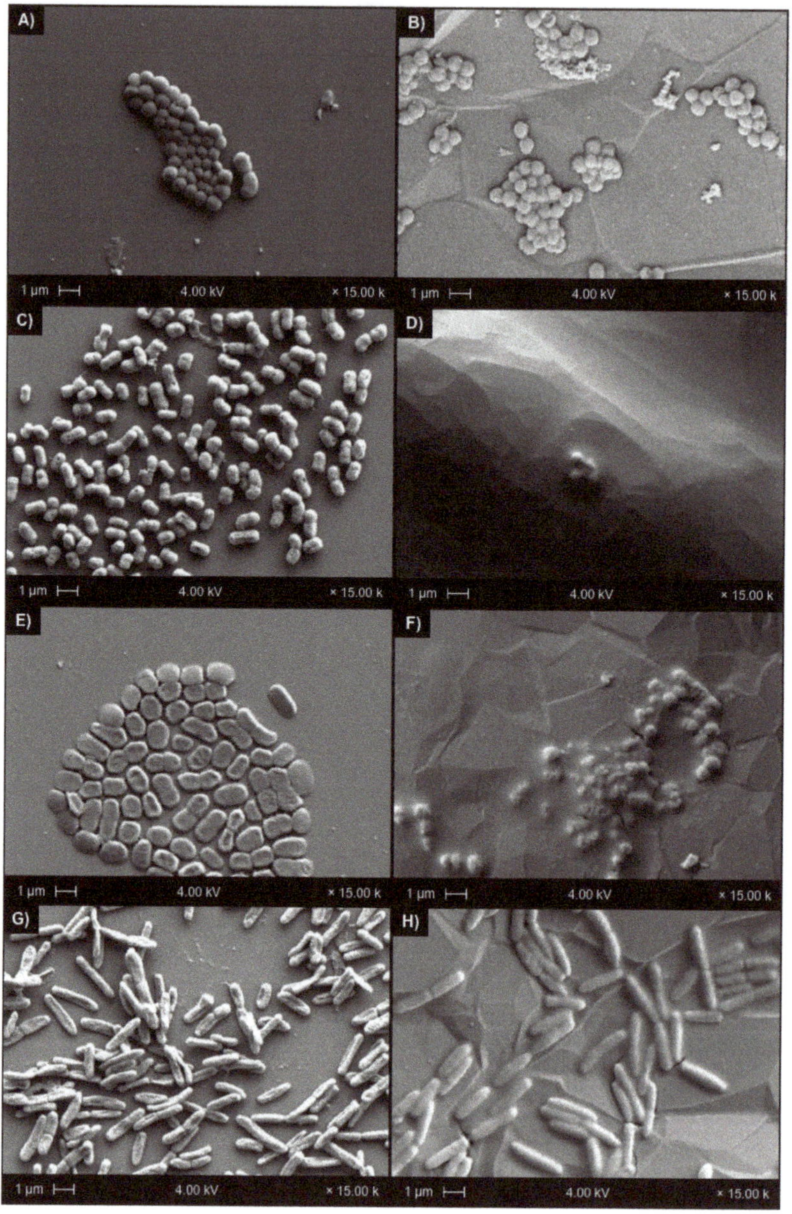

Figure 6. Comparison of the bacteria on (**A–D**) silicon and (**E–H**) graphene foam demonstrating the changes in the bacterial morphology against the resistant strains (**A**,**E**) *S. aureus* (**B**,**F**) *K. pneumoniae* (**C**,**G**) *A. baumannii* (**D**,**H**) *P. aeruginosa*.

4. Discussion

The emergence of AMR has resulted in enhanced morbidity and mortality rates, which in turn results in longer hospital stays, thus placing a financial burden on healthcare services worldwide [34–36]. There is currently a lack of effective therapies to combat AMR infections [4], if this is not addressed it has been proposed that that by 2050, AMR infection will supersede cancer as the leading cause of mortality worldwide [37]. One potential avenue to be explored is the use of metal ions as alternative therapeutic options [19,36,38,39]. The utilization of graphene has grown rapidly in the last decade due to the unique properties of graphene and its derivatives, such materials have been widely explored for use as potential biomaterials [40–43]. Throughout this study, six metal ions, silver, gold, palladium, platinum, zinc and gallium were evaluated against susceptible and resistant bacterial strains, *S. aureus, A. baumannii, K. pneumoniae* and *P. aeruginosa*. Furthermore, the metal ions were tested individually and as part of a coated graphene combination to determine if this had an effect on the antimicrobial activity observed.

The bacterial strains were firstly evaluated in regard to their antibiotic resistance profiles and categorized as either susceptible or resistant in order to determine if the metal ions could be utilized to eradicate AMR bacterial strains. Antimicrobial resistant bacterial strains are becoming increasingly difficult to treat with some current treatment options become largely ineffective. In a previous study conducted by Vaidya et al., (2017), disc diffusion assays were conducted using silver, platinum, gold and palladium ions against *K. pneumoniae* strain NCTC 9633 and *A. baumannii* strain NCTC 12156 [44]. In contrast to the work presented in this previous study, the results in the current study showed some differences in antimicrobial efficacy, this may be due to the different growth media and conditions utilised [44,45]. However, unlike Vaidya et al., the current study incorporated ionic metals with graphene foam in order to determine if this potentiated the antimicrobial activity. When tested alone, gold and palladium ions produced the greatest growth inhibition, although when tested in combination with the graphene foam (where the graphene is coated with the metal ions via a drop cast method), silver and platinum ions were as effective as gold and palladium. The combination of graphene foam with palladium ions resulted in synergistic antimicrobial activity against the resistant bacterial strains.

The antimicrobial properties of metals have been used throughout medical history [46]. The use of metals as antimicrobial agents declined after the discovery of antibiotics. However, due to the emergence of AMR there is a lack of effective therapeutic options [47]. Therefore, there has been a revival in the application of metals as biocidal agents [13,48]. The antimicrobial action of metals begins with the affinity of metal ions towards cellular components and biomolecules, which form stable complexes and damage vital bacterial cell processes [49]. Metals demonstrate toxic effects in several ways, such as displacement/damage of essential enzymes, blocking vital biomolecule functional groups or participating in cellular chemical reactions [13]. One or more of these processes may damage proteins, denature DNA, induce oxidative stress or affect the biological walls/membranes [13]. Usually, metals demonstrate more than one antimicrobial mechanism of action, simultaneously. Such antimicrobial mechanisms include, generation of ROS, the depletion of antioxidants, the metal ions can also bind to the thiol groups of DNA, enzymes and proteins [13,50–52]; which results in cellular membrane damage, disruption of electron transport and inhibit nutrient acquisition—resulting in cell lysis and ultimately death [53,54].

The antimicrobial activity of graphene and its derivatives remains controversial and further research is required to elucidate the antimicrobial mechanisms of action [23]. Previous studies suggest that the antimicrobial action of graphene extends to multi-drug resistant bacterial strains [38,55]. Karaky et al., (2020) previously determined that graphene (MIC: 125 μg mL^{-1}) and graphene oxide (MIC: >500 μg mL^{-1}) procured from the same source (Graphene Supermarket, USA) demonstrated little antimicrobial activity against a range of *P. aeruginosa* hospital isolates [38]. In a previous study, the antimicrobial activity of graphene was visualized using SEM, the presence of graphene sheets resulted in the loss of cell wall integrity in *E. coli* cells, resulting in cell death [56]. In the current study, SEM analysis revealed similar cellular ultrastructure damage to *S. aureus*, *K. pneumoniae* and *A. baumannii*.

The graphene foam used throughout this study has previously been characterised by Brownson et al., 2013 [57]. Raman spectroscopy revealed two characteristic peaks at ca. 1581 and 2684 cm^{-1} which are due to the G and 2D (G′) bands respectively; the highly symmetrical 2D band and the intensity ratio of the G to 2D band indicate that the surface is comprised of mono- to few-layer graphene sheets [57]. Furthermore, X-ray photoelectron spectroscopy (XPS) revealed the graphene foam comprised 95 % carbon and 5 % oxygen, whilst contact angle measurements revealed that the graphene foams exhibited a value of 120°, which indicated quasi-super-hydrophobicity [57].

Ionic gold, palladium and platinum demonstrated the greatest antimicrobial activity against the susceptible and resistant strains. Ionic gold like ionic silver, has a broad-spectrum of antimicrobial activity, reacting with carboxyl-, phosphate- and amino-groups, whilst also reducing the activity of lactate dehydrogenase [58,59]. The antimicrobial effect of platinum ions has been well documented since the discovery of cisplatin [Pt(II) (NH3)2C12] by Barnett Rosenberg in 1965. Rosenburg et al., 1965 were interested in the effect of electromagnetic radiation on both bacterial and mammalian cells when *E. coli* was incubated in a growth chamber with a set of platinum electrodes (which were considered inert), the electrical field resulted in morphological changes to the *E. coli* [14,60–62]. Palladium ions (Pd^{2+}) can inhibit enzyme function in eukaryotic cells and are known to inhibit creatine kinase, succinate dehydrogenase and other essential enzymatic processes [63]. In 2014, it was determined that concentrations as low as ca. 10^{-9} M of Pd^{2+} ions resulted in the inhibition of *S. aureus* growth after 24 h exposure [64]. The results presented in the current study suggests the utilization of graphene foam in combination with specific metal ions may have the potential to be used as part of a surface coating or wound dressing. Further research is required to potentially develop metal ion/graphene foam surface coatings, such as cytotoxic effects on mammalian cells and an enhanced understanding towards the precise mechanisms of action of metal ions. However, the use of metal ions/graphene foams in combination could reduce the risk of infection and the development of antimicrobial resistance generation [13,65].

5. Conclusions

The antimicrobial activity of six metal ions, silver, gold, palladium, platinum, zinc and gallium were evaluated against susceptible and resistant bacterial strains, *S. aureus*, *A. baumannii*, *K. pneumoniae* and *P. aeruginosa*. Furthermore, the metal ions were tested individually and in the presence of graphene foam (the metals were impregnated into the graphene surface via a drop cast method) to determine if this potentiated the antimicrobial activity observed. Ionic gold, palladium and platinum demonstrated the greatest antimicrobial activity against the susceptible and resistant strains. The combination of graphene foam with palladium ions resulted in synergistic antimicrobial activity against the resistant bacterial strains. Scanning electron microscopy (SEM) showed damage to the cellular ultrastructure when the bacterial strains were incubated upon the graphene foam alone. This study suggests that specific metal ions used in combination with graphene foam against specific bacterial strains (with different antibiotic-resistant profiles), could present a potential synergistic therapeutic option to treat AMR bacteria. Such combinations could potentially be used in a multitude of applications for example, wound dressings or surface coatings, effectively reducing wound infection and the transmission of common healthcare-acquired infections (HAIs).

Supplementary Materials: The following are available online at https://www.mdpi.com/2079-6412/11/3/352/s1, Table S1: Clinical Breakpoints determined by disc diffusion assays, data obtained via The European Committee on Antimicrobial Susceptibility Testing (EUCAST) valid from 1 January 2021 [33]. Antibiotics not classified by EUCAST have been omitted from this table. Abbreviations: ND; not determined.

Author Contributions: K.A.W. conceptualised the project. A.J.S., N.K., G.S.C. and J.A.B. were involved in data acquisition and analysis. C.E.B., A.J.M. and K.A.W. supervised the project and supplied the relevant materials. A.J.S. drafted the final manuscript. All authors have read and agreed to the published version of the manuscript.

Funding: This work was funded by a Health Research Accelerator Award courtesy of the University of Manchester.

Institutional Review Board Statement: Not applicable.

Informed Consent Statement: Not applicable.

Data Availability Statement: The datasets generated during the current study are available from the corresponding author on reasonable request.

Conflicts of Interest: The authors declare no conflict of interest.

References

1. Lowy, F.D. Antimicrobial resistance: The example of Staphylococcus aureus. *J. Clin. Investig.* **2003**, *111*, 1265–1273. [CrossRef] [PubMed]
2. Breathnach, A.S. Nosocomial infections and infection control. *Medicine* **2013**, *41*, 649–653. [CrossRef]
3. Bassetti, M.; Merelli, M.; Temperoni, C.; Astilean, A. New antibiotics for bad bugs: Where are we? *Ann. Clin. Microbiol. Antimicrob.* **2013**, *12*, 22. [CrossRef]
4. Ventola, C.L. The antibiotic resistance crisis: Part 1: Causes and threats. *Pharm. Ther.* **2015**, *40*, 277–283.
5. Gottrup, F. A specialized wound-healing center concept: Importance of a multidisciplinary department structure and surgical treatment facilities in the treatment of chronic wounds. *Am. J. Surg.* **2004**, *187*, S38–S43. [CrossRef]
6. Sydnor, E.R.M.; Perl, T.M. Hospital Epidemiology and Infection Control in Acute-Care Settings. *Clin. Microbiol. Rev.* **2011**, *24*, 141–173. [CrossRef]
7. Church, D.; Elsayed, S.; Reid, O.; Winston, B.; Lindsay, R. Burn Wound Infections. *Clin. Microbiol. Rev.* **2006**, *19*, 403–434. [CrossRef]
8. Liu, X.; Lee, P.-Y.; Ho, C.-M.; Lui, V.C.H.; Chen, Y.; Che, C.-M.; Tam, P.K.H.; Wong, K.K.Y. Silver Nanoparticles Mediate Differential Responses in Keratinocytes and Fibroblasts during Skin Wound Healing. *ChemMedChem* **2010**, *5*, 468–475. [CrossRef]
9. Lu, L.-C.; Chen-Wen, L.; Qing, W.; Min, H.; San-Jun, S.; Zi-Wei, L.; Guo-Lin, W.; Huan-Huan, C.; Yuan-Yuan, L.; Qian, Z.; et al. Silver nanoparticles/chitosan oligosaccharide/poly(vinyl alcohol) nanofiber promotes wound healing by activating TGFβ1/Smad signaling pathway. *Int. J. Nanomed.* **2016**, *11*, 373–387. [CrossRef] [PubMed]
10. McQuitty, R.J. Metal-based Drugs. *Sci. Prog.* **2014**, *97*, 1–19. [CrossRef] [PubMed]
11. Southam, H.M.; Butler, J.A.; Chapman, J.A.; Poole, R.K. The Microbiology of Ruthenium Complexes. In *Advances in Microbial Physiology*; Elsevier: Amsterdam, The Netherlands, 2017; Volume 71, pp. 1–96. [CrossRef]
12. Zhou, Y.; Kong, Y.; Kundu, S.; Cirillo, J.D.; Liang, H. Antibacterial activities of gold and silver nanoparticles against Escherichia coli and bacillus Calmette-Guérin. *J. Nanobiotechnol.* **2012**, *10*, 19. [CrossRef] [PubMed]
13. Lemire, J.A.; Harrison, J.J.; Turner, R.J. Antimicrobial activity of metals: Mechanisms, molecular targets and applications. *Nat. Rev. Genet.* **2013**, *11*, 371–384. [CrossRef]
14. Slate, A.J.; Whitehead, K.A.; Brownson, D.A.; Banks, C.E. Microbial fuel cells: An overview of current technology. *Renew. Sustain. Energy Rev.* **2019**, *101*, 60–81. [CrossRef]
15. Bianco, A.; Cheng, H.-M.; Enoki, T.; Gogotsi, Y.; Hurt, R.H.; Koratkar, N.; Kyotani, T.; Monthioux, M.; Park, C.R.; Tascon, J.M.; et al. All in the graphene family—A recommended nomenclature for two-dimensional carbon materials. *Carbon* **2013**, *65*, 1–6. [CrossRef]
16. Novoselov, K.S.; Geim, A.K.; Morozov, S.V.; Jiang, D.; Zhang, Y.; Dubonos, S.V.; Grigorieva, I.V.; Firsov, A.A. Electric Field Effect in Atomically Thin Carbon Films. *Science* **2004**, *306*, 666–669. [CrossRef] [PubMed]
17. Novoselov, K.S.; Fal'ko, V.I.; Colombo, L.; Gellert, P.R.; Schwab, M.G.; Kim, K. A roadmap for graphene. *Nature* **2012**, *490*, 192–200. [CrossRef]
18. Shao, Y.; Wang, J.; Wu, H.; Liu, J.; Aksay, I.A.; Lin, Y. Graphene Based Electrochemical Sensors and Biosensors: A Review. *Electroanalysis* **2010**, *22*, 1027–1036. [CrossRef]
19. Slate, A.; Karaky, N.; Whitehead, K.A. Antimicrobial properties of Modified Graphene and other advanced 2D Material Coated Surfaces. In *2D Materials—Characterization, Production and Applications*; Banks, C.E., Brownson, D.A.C., Eds.; CRC Press: Boca Raton, FL, USA, 2018; pp. 86–104. [CrossRef]
20. Choi, W.; Lahiri, I.; Seelaboyina, R.; Kang, Y.S. Synthesis of Graphene and Its Applications: A Review. *Crit. Rev. Solid State Mater. Sci.* **2010**, *35*, 52–71. [CrossRef]
21. Liu, J.; Cui, L.; Losic, D. Graphene and graphene oxide as new nanocarriers for drug delivery applications. *Acta Biomater.* **2013**, *9*, 9243–9257. [CrossRef]

22. Zou, X.; Zhang, L.; Wang, Z.; Luo, Y. Mechanisms of the Antimicrobial Activities of Graphene Materials. *J. Am. Chem. Soc.* **2016**, *138*, 2064–2077. [CrossRef]
23. Hegab, H.M.; Elmekawy, A.; Zou, L.; Mulcahy, D.; Saint, C.P.; Ginic-Markovic, M. The controversial antibacterial activity of graphene-based materials. *Carbon* **2016**, *105*, 362–376. [CrossRef]
24. Chen, J.; Peng, H.; Wang, X.; Shao, F.; Yuan, Z.; Han, H. Graphene oxide exhibits broad-spectrum antimicrobial activity against bacterial phytopathogens and fungal conidia by intertwining and membrane perturbation. *Nanoscale* **2014**, *6*, 1879–1889. [CrossRef]
25. Akhavan, O.; Ghaderi, E. Toxicity of Graphene and Graphene Oxide Nanowalls Against Bacteria. *ACS Nano* **2010**, *4*, 5731–5736. [CrossRef] [PubMed]
26. Carpio, I.E.M.; Santos, C.M.; Wei, X.; Rodrigues, D.F. Toxicity of a polymer–graphene oxide composite against bacterial planktonic cells, biofilms, and mammalian cells. *Nanoscale* **2012**, *4*, 4746–4756. [CrossRef] [PubMed]
27. Dallavalle, M.; Calvaresi, M.; Bottoni, A.; Melle-Franco, M.; Zerbetto, F. Graphene Can Wreak Havoc with Cell Membranes. *ACS Appl. Mater. Interfaces* **2015**, *7*, 4406–4414. [CrossRef] [PubMed]
28. Butler, J.A.; Osborne, L.; El Mohtadi, M.; Whitehead, K.A. Graphene derivatives potentiate the activity of antibiotics against *Enterococcus faecium*, *Klebsiella pneumoniae* and *Escherichia coli*. *AIMS Environ. Sci.* **2020**, *7*, 106–113. [CrossRef]
29. Gill, E.E.; Franco, O.L.; Hancock, R.E.W. Antibiotic Adjuvants: Diverse Strategies for Controlling Drug-Resistant Pathogens. *Chem. Biol. Drug Des.* **2014**, *85*, 56–78. [CrossRef]
30. Ling, L.L.; Schneider, T.; Peoples, A.J.; Spoering, A.L.; Engels, I.; Conlon, B.P.; Mueller, A.; Schäberle, T.F.; Hughes, D.E.; Epstein, S.S.; et al. A new antibiotic kills pathogens without detectable resistance. *Nature* **2015**, *517*, 455–459. [CrossRef]
31. Picconi, P.; Hind, C.K.; Nahar, K.S.; Jamshidi, S.; Di Maggio, L.; Saeed, N.; Evans, B.; Solomons, J.; Wand, M.E.; Sutton, J.M.; et al. New Broad-Spectrum Antibiotics Containing a Pyrrolobenzodiazepine Ring with Activity against Multidrug-Resistant Gram-Negative Bacteria. *J. Med. Chem.* **2020**, *63*, 6941–6958. [CrossRef] [PubMed]
32. Butler, J.A.; Slate, A.J.; Todd, D.B.; Airton, D.; Hardman, M.; Hickey, N.A.; Scott, K.; Venkatraman, P.D. A traditional Ugandan Ficus natalensis bark cloth exhibits antimicrobial activity against methicillin-resistant Staphylococcus aureus. *J. Appl. Microbiol.* 2020. [CrossRef]
33. The European Committee on Antimicrobial Susceptibility Testing. Breakpoint Tables for Interpretation of MICs and Zone Diameters, Version 11.0. 2021. Available online: https://eucast.org/clinical_breakpoints/ (accessed on 29 January 2021).
34. Llor, C.; Bjerrum, L. Antimicrobial resistance: Risk associated with antibiotic overuse and initiatives to reduce the problem. *Ther. Adv. Drug Saf.* **2014**, *5*, 229–241. [CrossRef]
35. Blair, J.M.A.; Webber, M.A.; Baylay, A.J.; Ogbolu, D.O.; Piddock, L.J.V. Molecular mechanisms of antibiotic resistance. *Nat. Rev. Genet.* **2014**, *13*, 42–51. [CrossRef]
36. Slate, A.J.; Shalamanova, L.; Akhidime, I.D.; Whitehead, K.A. Rhenium and yttrium ions as antimicrobial agents against multidrug resistant *Klebsiella pneumoniae* and *Acinetobacter baumannii* biofilms. *Lett. Appl. Microbiol.* **2019**, *69*, 168–174. [CrossRef] [PubMed]
37. O'Neill, J. *Tackling Drug-Resistant Infections Globally: Final Report and Recommendations*; HM Government and Welcome Trust: London, UK, 2016.
38. Karaky, N.; Kirby, A.; McBain, A.J.; Butler, J.A.; El Mohtadi, M.; Banks, C.E.; Whitehead, K.A. Metal ions and graphene-based compounds as alternative treatment options for burn wounds infected by antibiotic-resistant Pseudomonas aeruginosa. *Arch. Microbiol.* **2020**, *202*, 995–1004. [CrossRef] [PubMed]
39. Wahid, F.; Zhong, C.; Wang, H.-S.; Hu, X.-H.; Chu, L.-Q. Recent Advances in Antimicrobial Hydrogels Containing Metal Ions and Metals/Metal Oxide Nanoparticles. *Polymers* **2017**, *9*, 636. [CrossRef] [PubMed]
40. Ege, D.; Kamali, A.R.; Boccaccini, A.R. Graphene Oxide/Polymer-Based Biomaterials. *Adv. Eng. Mater.* **2017**, *19*. [CrossRef]
41. Feng, L.; Liu, Z. Graphene in biomedicine: Opportunities and challenges. *Nanomedicine* **2011**, *6*, 317–324. [CrossRef]
42. Han, S.; Sun, J.; He, S.; Tang, M.; Chai, R. The application of graphene-based biomaterials in biomedicine. *Am. J. Transl. Res.* **2019**, *11*, 3246–3260. [PubMed]
43. Reddy, S.; Xu, X.; Guo, T.; Zhu, R.; He, L.; Ramakrishana, S. Allotropic carbon (graphene oxide and reduced graphene oxide) based biomaterials for neural regeneration. *Curr. Opin. Biomed. Eng.* **2018**, *6*, 120–129. [CrossRef]
44. Vaidya, M.Y.; McBain, A.J.; Butler, J.A.; Banks, C.E.; Whitehead, K.A. Antimicrobial Efficacy and Synergy of Metal Ions against *Enterococcus faecium*, *Klebsiella pneumoniae* and *Acinetobacter baumannii* in Planktonic and Biofilm Phenotypes. *Sci. Rep.* **2017**, *7*, 1–9. [CrossRef]
45. Verran, J.; Redfern, J.; Smith, L.A.; Whitehead, K.A. A critical evaluation of sampling methods used for assessing microorganisms on surfaces. *Food Bioprod. Process.* **2010**, *88*, 335–340. [CrossRef]
46. Elsome, A.M.; Hamilton-Miller, J.M.T.; Brumfitt, W.; Noble, W.C. Antimicrobial activities in vitro and in vivo of transition element complexes containing gold(I) and osmium(VI). *J. Antimicrob. Chemother.* **1996**, *37*, 911–918. [CrossRef] [PubMed]
47. Gold, K.; Slay, B.; Knackstedt, M.; Gaharwar, A.K. Antimicrobial Activity of Metal and Metal-Oxide Based Nanoparticles. *Adv. Ther.* **2018**, *1*. [CrossRef]
48. Dizaj, S.M.; Lotfipour, F.; Barzegar-Jalali, M.; Zarrintan, M.H.; Adibkia, K. Antimicrobial activity of the metals and metal oxide nanoparticles. *Mater. Sci. Eng. C* **2014**, *44*, 278–284. [CrossRef]
49. Nies, D.H. Microbial heavy-metal resistance. *Appl. Microbiol. Biotechnol.* **1999**, *51*, 730–750. [CrossRef] [PubMed]

50. Feng, Q.L.; Wu, J.; Chen, G.Q.; Cui, F.; Kim, T.; Kim, J. A mechanistic study of the antibacterial effect of silver ions on *Escherichia coli* and *Staphylococcus aureus*. *J. Biomed. Mater. Res.* **2000**, *52*, 662–668. [CrossRef]
51. Stiefel, P.; Schmidt-Emrich, S.; Maniura-Weber, K.; Ren, Q. Critical aspects of using bacterial cell viability assays with the fluorophores SYTO9 and propidium iodide. *BMC Microbiol.* **2015**, *15*, 36–39. [CrossRef]
52. Kim, J.S.; Kuk, E.; Yu, K.N.; Kim, J.-H.; Park, S.J.; Lee, H.J.; Kim, S.H.; Park, Y.K.; Park, Y.H.; Hwang, C.-Y.; et al. Antimicrobial effects of silver nanoparticles. *Nanomed. Nanotechnol. Biol. Med.* **2007**, *3*, 95–101. [CrossRef]
53. Sondi, I.; Salopek-Sondi, B. Silver nanoparticles as antimicrobial agent: A case study on E. coli as a model for Gram-negative bacteria. *J. Colloid Interface Sci.* **2004**, *275*, 177–182. [CrossRef] [PubMed]
54. Hobman, J.L.; Crossman, L.C. Bacterial antimicrobial metal ion resistance. *J. Med. Microbiol.* **2015**, *64*, 471–497. [CrossRef] [PubMed]
55. Shoeb, M.; Mobin, M.; Rauf, M.A.; Owais, M.; Naqvi, A.H. In Vitro and in Vivo Antimicrobial Evaluation of Graphene–Polyindole (Gr@PIn) Nanocomposite against Methicillin-Resistant *Staphylococcus aureus* Pathogen. *ACS Omega* **2018**, *3*, 9431–9440. [CrossRef] [PubMed]
56. Oh, H.G.; Lee, J.-Y.; Son, H.G.; Kim, D.H.; Park, S.-H.; Kim, C.M.; Jhee, K.-H.; Song, K.S. Antibacterial mechanisms of nanocrystalline diamond film and graphene sheet. *Results Phys.* **2019**, *12*, 2129–2135. [CrossRef]
57. Brownson, D.A.C.; Figueiredo-Filho, L.C.S.; Ji, X.; Gómez-Mingot, M.; Iniesta, J.; Fatibello-Filho, O.; Kampouris, D.K.; Banks, C.E. Freestanding three-dimensional graphene foam gives rise to beneficial electrochemical signatures within non-aqueous media. *J. Mater. Chem. A* **2013**, *1*, 5962–5972. [CrossRef]
58. Shareena Dasari, T.P.; Zhang, Y.; Yu, H. Antibacterial Activity and Cytotoxicity of Gold (I) and (III) Ions and Gold Nanoparticles. *Biochem. Pharmacol. Open Access* **2015**, *4*, 1–5. [CrossRef]
59. Jung, W.K.; Koo, H.C.; Kim, K.W.; Shin, S.; Kim, S.H.; Park, Y.H. Antibacterial Activity and Mechanism of Action of the Silver Ion in *Staphylococcus aureus* and *Escherichia coli*. *Appl. Environ. Microbiol.* **2008**, *74*, 2171–2178. [CrossRef]
60. Tylkowski, B.; Jastrząb, R.; Odani, A. Developments in platinum anticancer drugs. *Phys. Sci. Rev.* **2018**, *3*. [CrossRef]
61. Monneret, C. Platinum anticancer drugs. From serendipity to rational design. *Ann. Pharm. Fr.* **2011**, *69*, 286–295. [CrossRef]
62. Rosenberg, B.; Van Camp, L.; Krigas, T. Inhibition of Cell Division in Escherichia coli by Electrolysis Products from a Platinum Electrode. *Nature* **1965**, *205*, 698–699. [CrossRef]
63. Liu, T.Z.; Lee, S.D.; Bhatnagar, R.S. Toxicity of palladium. *Toxicol. Lett.* **1979**, *4*, 469–473. [CrossRef]
64. Adams, C.P.; Walker, K.A.; Obare, S.O.; Docherty, K.M. Size-Dependent Antimicrobial Effects of Novel Palladium Nanoparticles. *PLoS ONE* **2014**, *9*, e85981. [CrossRef] [PubMed]
65. Richtera, L.; Chudobova, D.; Cihalova, K.; Kremplova, M.; Milosavljevic, V.; Kopel, P.; Blazkova, I.; Hynek, D.; Adam, V.; Kizek, R. The Composites of Graphene Oxide with Metal or Semimetal Nanoparticles and Their Effect on Pathogenic Microorganisms. *Materials* **2015**, *8*, 2994–3011. [CrossRef]

coatings

MDPI

Article

Sputter-Deposited Ag Nanoparticles on Electrospun PCL Scaffolds: Morphology, Wettability and Antibacterial Activity

Daniele Valerini [1,*], Loredana Tammaro [2,*], Roberta Vitali [3,*], Gloria Guillot [4] and Antonio Rinaldi [5,*]

[1] SSPT-PROMAS-MATAS, ENEA—Italian National Agency for New Technologies, Energy and Sustainable Economic Development, S.S. 7 Appia, km 706, 72100 Brindisi, Italy
[2] SSPT-PROMAS-NANO, ENEA—Italian National Agency for New Technologies, Energy and Sustainable Economic Development, Piazzale E. Fermi, 1, Portici, 80055 Napoli, Italy
[3] SSPT-TECS-TEB, ENEA—Italian National Agency for New Technologies, Energy and Sustainable Economic Development, Via Anguillarese 301, 00123 Rome, Italy
[4] NANOFABER srl, Via Anguillarese 301, 00123 Rome, Italy; gloria.guillot@nanofaber.com
[5] SSPT-PROMAS-MATPRO, ENEA—Italian National Agency for New Technologies, Energy and Sustainable Economic Development, Via Anguillarese 301, 00123 Rome, Italy
* Correspondence: daniele.valerini@enea.it (D.V.); loredana.tammaro@enea.it (L.T.); roberta.vitali@enea.it (R.V.); antonio.rinaldi@enea.it (A.R.)

Abstract: Porous scaffolds made of biocompatible and environmental-friendly polymer fibers with diameters in the nano/micro range can find applications in a wide variety of sectors, spanning from the biomedical field to textiles and so on. Their development has received a boost in the last decades thanks to advances in the production methods, such as the electrospinning technique. Conferring antimicrobial properties to these fibrous structures is a primary requirement for many of their applications, but the addition of antimicrobial agents by wet methods can present a series of drawbacks. In this work, strong antibacterial action is successfully provided to electrospun polycaprolactone (PCL) scaffolds by silver (Ag) addition through a simple and flexible way, namely the sputtering deposition of silver onto the PCL fibers. SEM-EDS analyses demonstrate that the polymer fibers get coated by Ag nanoparticles without undergoing any alteration of their morphological integrity upon the deposition process. The influence on wettability is evaluated with polar (water) and non-polar (diiodomethane) liquids, evidencing that this coating method allows preserving the hydrophobic character of the PCL polymer. Excellent antibacterial action (reduction > 99.995% in 4 h) is demonstrated against *Escherichia coli*. The easy fabrication of these PCL-Ag mats can be applicable to the production of biomedical devices, bioremediation and antifouling systems in filtration, personal protective equipment (PPE), food packaging materials, etc.

Keywords: antimicrobial; polycaprolactone (PCL); silver nanoparticles; nanofibers; electrospinning; sputtering; antibacterial; antiviral; biomedical; bioremediation; antifouling

Citation: Valerini, D.; Tammaro, L.; Vitali, R.; Guillot, G.; Rinaldi, A. Sputter-Deposited Ag Nanoparticles on Electrospun PCL Scaffolds: Morphology, Wettability and Antibacterial Activity. *Coatings* **2021**, *11*, 345. https://doi.org/10.3390/coatings11030345

Academic Editor: Jean-François Berret

Received: 23 February 2021
Accepted: 12 March 2021
Published: 18 March 2021

1. Introduction

In the last decade, the development of the electrospinning technique has enabled the fabrication of an unprecedented variety of fibrous scaffolds from synthetic and natural materials. Electrospinning is a polymer processing technique of primary interest in scaffold fabrication due to its ability to seamlessly produce fibers with diameters in the range from micrometers down to tens of nanometers, high surface area to volume ratio structure, high porosity and capability to mimic the extracellular matrix (ECM), needful requirements in several applications such as medicine, filtration, textiles, etc. [1–5].

The desirable characteristics of an ideal scaffold are highly dependent on the used polymer and on the given application. Among polymeric materials, Poly(ε-caprolactone) (PCL) is a linear aliphatic polyester with good biocompatibility, ECM mimicking possibility, bioresorbability, biodegradability, thermal stability, mechanical strength, elasticity, non-toxicity, slow degradability and low cost, which are desirable characteristics for tissue engineering applications, food packaging, drug delivery systems, wound dressings, filtration and antibacterial constructs [5–8]. However, PCL-derived materials have no antimicrobial properties, unless further functionalized, for example with silver.

Silver (Ag) is one of the most used antimicrobial materials, with biocide action reported against a wide range of bacterial species as well as viruses, so that Ag nanoparticles (NPs) are often added in composite materials to confer them antibacterial properties [9]. Ag-functionalized PCL films are usually produced from a starting solution containing both Ag NPs and PCL mixed together, which is then processed by casting [10,11] or electrospinning [12–14] methods. In spite of the convenience of such procedures, they present some drawbacks. For example, the hydrophilicity of chemically synthesized Ag NPs and the hydrophobicity of PCL can result in composite films with non-uniform NPs dispersion and irregular ion release phenomena for certain applications [10,11]. Hence, in situ chemical routes are often employed to synthesize Ag NPs within the same deposition solution [11–13], which, however, can still present many inconveniences, like: addition of a further step in the process, more difficult electrospinning process optimization with the biphasic system (Ag precursor and PCL) with respect to the simpler PCL only, use of toxic and corrosive Ag precursors, control of Ag concentration that is conditional on the complete precursor reduction, and use of additional co-solvents to avoid precipitation and favor chemical reaction [11–13]. Additionally, the hydrophilic Ag NPs in the host polymer matrix can shift the PCL hydrophobic character towards a more hydrophilic one [11,14], which could be undesirable under certain circumstances. Finally, starting from a mixed blend of PCL and Ag, only a limited amount of silver is directly exposed on the fiber surfaces and it is not possible to achieve selective functionalization of some desired portions of the membrane only.

As an alternative method, electrospun PCL mats can be post-processed through subsequent Ag addition by dip coating in silver nitrate aqueous solution, followed by drying and UV photoreduction [15]. Yet, this wet route still presents issues related to wettability mismatches (as observed in [15]), use of unsafe precursors, expected complete reduction of precursors, and impregnation of the whole mat with the Ag solution.

Our study addresses a different strategy, consisting in the addition of Ag nanoparticles onto electrospun bare PCL mats by the sputtering technique, which overcomes many of the drawbacks mentioned above and provides a modular manufacturing route of broader applicability, thus representing a simple and flexible way to functionalize the starting polymer scaffold. Sputtering is an eco-friendly, relatively cheap, widely commercially and industrially employed PVD (physical vapor deposition) method, allowing the deposition of several kinds of materials even on large areas. By properly setting the deposition parameters, sputtering is capable to deposit antibacterial coatings even on delicate materials like soft polymers and bioplastics without significantly damaging them [16]. Sputter deposition of antimicrobial materials, also combining multiple elements together, allows the production of uniform and adherent nanostructured coatings, reducing possible release of particles harmful to human health, and permitting easy and accurate control of the amount of deposited material by tuning the process parameters, like deposition time or power [17,18]. Of course, being decoupled from the initial PCL electrospinning process, the sputtering step can be even applied on commercial PCL mats and third-party substrates. Additionally, the physical process of Ag NPs sputter-deposition can confer a slightly hydrophobic character to the treated substrate, as already shown in a previous paper on a different polymer substrate [19], being useful when needing to retain the hydrophobicity of the PCL surface. As a further advantage our method could provide Ag NPs deposited onto the fiber surfaces, promoting the direct exposure of the active antibacterial agent to the surrounding environment. Finally, Ag NPs are deposited in the shallow regions rather than deep within the mat and, by proper masking, they can be placed on selected areas only, thus avoiding deposition/contamination where no functionalization is required.

So far, only few works reported the functionalization of electrospun PCL-containing scaffolds by the Ag sputtering process [20,21]. However, in those works poly(glycerol sebacate) (PGS) was used as base polymer, while PCL was added to allow the PGS electrospinnability, and their main focus was on the electrical and thermomechanical properties. Only the second report [21] presented some antibacterial results, through the disk diffusion method, on PGS/PCL coated with Ag films with thickness from 50 to 275 nm.

Hence, in the present work, we functionalized the surface of electrospun PCL fibers by sputtering a nanoscale silver coating in the form of Ag nanoparticles, endowing the polymer material with marked antibacterial activity despite a very low Ag content. The results presented hereafter demonstrate the production of PCL-Ag mats through an easy, flexible and effective way, avoiding the disadvantages affecting other methods described above, with possible applications in biomedical devices, food packaging, and sustainability applications like bioremediation and antifouling coatings in filtration. In particular, this class of PCL electrospun sheets have been previously demonstrated to have potential for bioremediation and the production of industrial enzymes and organic acids [22,23]. The first report [22] demonstrated that bare (i.e., uncoated) PCL meshes are highly susceptible to be colonized by pathogenic relevant bacteria, which can be exploited in chronic wound environment to remove biofilms from the wound, thus helping with cleaning and disinfection. In the second study [23], the authors pointed out the satisfactory resistance of these PCL sheets to microbial degradation (compared to polylactic acid) and their ability to support a dense biofilm of pure *S. fuliginis* (former *Flavobacterium* sp. ATCC 27551), a very relevant strain in bioremediation of organophosphorus compounds. However, those authors concluded that a functionalization of such PCL sheets would be needed to properly modulate or inhibit microbial activity in many applications, such antifouling surfaces. The coating treatment proposed in this paper also addresses this need.

2. Materials and Methods

2.1. Electrospinning of PCL Scaffolds

PCL microfibrous sheets with a thickness of 150 μm were produced via electrospinning by company Nanofaber srl (Rome, Italy) using a protocol derived by the commercially available grade (NBARE™ series) to increase the percentage of submicrometer fibers. Process parameters for this batch are reported in Table 1. The sheets were fabricated using a standard needle-technology electrospinning equipment (Fluidnatek LE100, Bioinicia, Spain) outfitted with a flat collector and a two axes emitter motion to process at room temperature a 12% *w*/*v* solution of PCL polymer dissolved in a mixture solvent of DMF/chloroform 2:8. To prepare the electrospinning solution, pure PCL granules of CAPA® 6800 (80,000 MW, Perstorp, Sweden) were stirred as long as needed for complete dissolution in said solvent made of dimethylformamide (100% purity, VWR, Radnor, PA, USA) and chloroform (99.2% purity, stabilized with 0.6% ethanol, VWR, Radnor, PA, USA).

Table 1. Process parameters used for the electrospinning of PCL solution 12% *w/v*.

Flow Rate (μL/h)	Applied Voltage (kV)	Needle Inner Diameter (mm)	Working Distance (cm)	Deposition Time (min)
4000	25	1.8	17	30

Finally, the PCL sheets were mechanically die-cut into disks of 15 mm diameter to fit in the multiwell plates used for the biological assessment.

2.2. Sputter-Deposition of Ag Nanoparticles

The electrospun PCL disks, together with reference silicon substrates, were placed in the vacuum chamber evacuated to a base pressure of 1.5×10^{-4} Pa for Ag deposition by rf magnetron sputtering. The deposition was carried out at room temperature in Ar atmosphere at working pressure of 3 Pa. Ag target (purity 99.99%, diameter 10 cm) was sputtered by a low power of 50 W to avoid damaging the PCL fiber bundles with the deposition bombardment. The substrates were rotated by a planetary rotation system at 5 rpm, passing under the target plasma for 11 cycles. The nominal thickness of silver deposited on reference Si substrate was 6 nm, which, considering the substrate coverage deduced from SEM inspection, corresponds to Ag load of about 4.4 $\mu g/cm^2$.

2.3. Scanning Electron Microscopy (SEM) and Energy Dispersive X-ray Spectroscopy (EDS)

The morphological properties of PCL electrospun meshes were examined with a field emission gun scanning electron microscope Leo 1530 model (ZEISS, Jena, Germany) working at low voltage (2 kV) to avoid charging effects and damage to the dielectric PCL polymer from overheating. Sheet thickness was also measured via SEM on cross-section samples.

Fiber distribution was obtained by measuring the diameter of fibers intersecting a grid sufficiently spaced to accommodate at least 50 intercepting fibers. The SEM image was pre-processed using "ImageJ" software (Rasband, W.S., ImageJ, U.S. National Institutes of Health, Bethesda, MD, USA) to convert images in black & white.

Energy dispersive X-ray spectroscopy (EDS) measurements were taken using a X-MAX detector (Oxford, UK). PCL sheets samples were examined to map the distribution of Ag on coated samples, however the operating voltage of the SEM was raised to 5 kV to observe Ag peaks in the EDS spectrum.

2.4. Porosity of PCL Scaffolds

Gravimetric technique was used for measuring the porosity of electrospun samples. Using a balance (ORMA, BCA120) with precision of 10^{-4} g, knowing the volume of the given sheet sample, the apparent density $\rho*$ of PCL sheets was calculated as:

$$\rho* = \text{measured mass/given volume} \tag{1}$$

The scaffold porosity ε was obtained using the formula:

$$\varepsilon = 1 - \rho*/\rho \tag{2}$$

upon assuming a value of ρ = 1.145 g/cm^3 for PCL. Porosity measurements were replicated at least three times for each type of samples, with reported ε being the average.

2.5. Contact Angle Measurements

Wetting tests were performed on a contact angle goniometer (OCA 20, Dataphysics, Filderstadt, Germany) operating at room temperature. The contact angle was measured by the sessile drop method using 1 μL droplet volume and deposition rate of 1 μL/s. Milli-Q water and diiodomethane were used as the polar and the apolar substances, respectively. The images were captured after 10, 60 and 120 s and data were collected with SCA 202 software (version 3.4.3 build 76). For statistical accountability, the average of ten contact angle measurements for each sample was calculated.

2.6. Antimicrobial Tests

Escherichia coli (*E. coli*) stock cultures kept at −80 °C in 10% (wt./vol.) glycerol were inoculated into 5 mL of Luria-Bertani (LB) broth and incubated at 37 °C O/N before their use in experiments. *E. coli* was pre-inoculated aerobically for 16 h at 37 °C in LB medium, with constant shaking at 250 rpm. The day of the test the bacteria were diluted and grown until OD600 was 0.025, about 5.5×10^6 colony forming units/mL (CFU/mL). Then the bacteria were placed in multiwell plate with 24-wells in the presence of uncoated PCL samples (blank disks) or Ag-coated samples and incubated at 37 °C under constant agitation at 50 rpm. A control with *E. coli* alone was also inserted. At following times t = 0, 2 and 4 h, 10 μL of each sample were diluted in PBS (serial dilution from 10^{-1} to 10^{-4}) and 10 μL of each dilution were distributed on LB agar dishes (15 g L^{-1} agar) and incubated for 18 h at 37 °C. Each plating was performed in triplicates. Subsequently, the number of CFU/mL was quantified for each sample. The experiment was repeated twice.

3. Results and Discussion

Figure 1 shows the picture of some representative samples of electrospun blank PCL disks as obtained from the electrospinning process, and PCL-Ag disks from the subsequent sputter-deposition of Ag on the electrospun PCL mat.

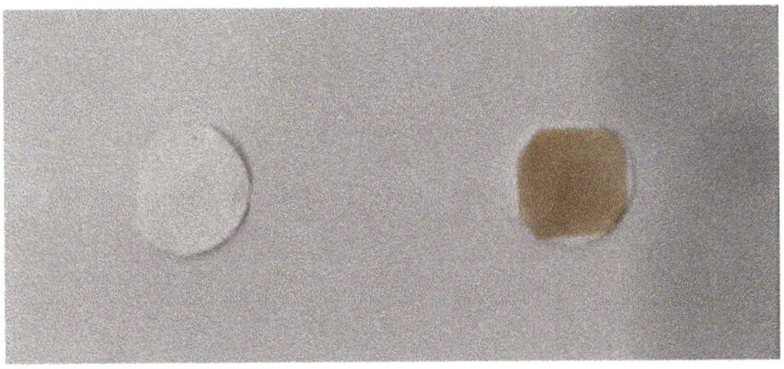

Figure 1. Picture of representative samples of electrospun blank PCL (**left**) and Ag-coated PCL (**right**).

3.1. SEM-EDS Analyses

The samples microstructure was characterized by SEM. The membrane, imaged at low magnification, appears macroscopically dense, as a thick maze of intertwined randomly oriented fibers. The diameter of the fibers is broadly distributed from 100 nm or less to some microns. The fiber diameter distribution as determined by fiber counting on SEM micrographs was 1.83 ± 0.54 µm and the average porosity 80.9 ± 1.8%. The morphology of the PCL sheets before and after the Ag treatment is displayed in the micrographs of Figure 2. It appears essentially identical apart from the higher contrast in the treated PCL, reflecting a higher SEM signal from the Ag metallization. Remarkably, this indicates that the proposed coating process is non-destructive and sufficiently gentle to be performed on a temperature sensitive polymer such as PCL (melting temperature about 50–60 °C).

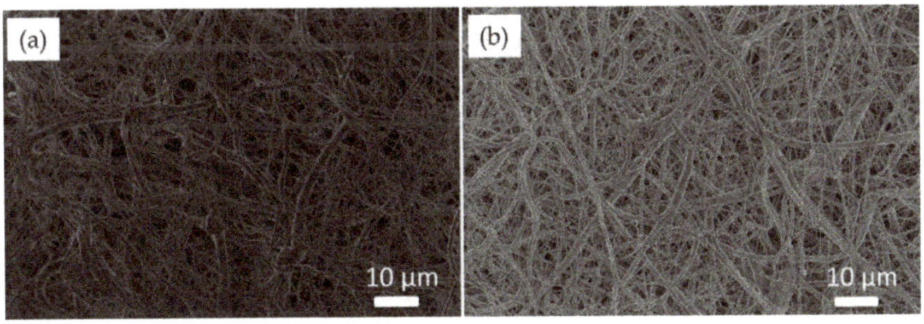

Figure 2. SEM micrographs at low voltage showing an overview at low magnification of (**a**) the bare sample vs. (**b**) Ag-coated one.

The close-up micrographs in Figure 3 further demonstrate that the coating process does not alter morphology and texture, while revealing the Ag particles decorating the PCL fibers. From the images of the coated fibers it was possible to identify Ag particles with lateral dimensions ranging from a few nanometers to some ten of nanometers, together with some bigger clusters up to around 80 nm. From particle counting on the SEM micrographs, the average nanoparticle diameter was determined to be (22 ± 9) nm, which, however, does not take into account particles smaller than about 9 nm, as they cannot be correctly measured due to image resolution at the given magnifications.

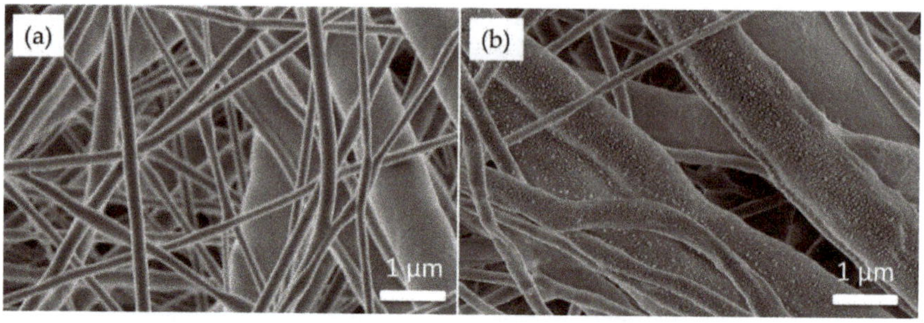

Figure 3. Close-up SEM micrographs at low voltage showing (**a**) the fibers in the bare sample vs. (**b**) fibers of the Ag-coated one, where silver decoration is clearly visible.

The absence of any evident damage (modifications in fibers shape or dimension, fiber fusion or collapse, occlusion of pores, etc.) in our conditions is also noteworthy when looking to analogous experimental findings for Ag deposition onto electrospun polymer fibers. In a recent paper [24], a different PVD process was employed to coat various polymer fibers, finding no damages on some polymers while important alterations and degradation on another. Such effects may depend on the polymeric substrate properties, the amount of silver deposited, as well as the characteristics of the deposition process (deposition technique and process parameters employed) that influence the energy of the depositing species and the consequent impact on the coated fibers. In our case, the suitable combination of deposition technique, process parameters and amount of deposited silver allowed the functionalization of the soft PCL substrate without inducing significant damages. At the same time, the kinetic energy of the sputtered Ag particles impinging on the soft polymer fibers is expected to promote a good adhesion of the silver nanocoating, preventing their premature detachment and release of large amount of particles that could result harmful for living organisms. Although these aspects need ad hoc analyses to get an exact evaluation, which could be object of future works, a preliminary rough hint on coating stability has been here deduced from the absence of any evident visual damage on the coated samples during their manipulation and after intentional manual bending and attempts to scratch them.

Since Ag was deposited as a nanoparticle layer around the fiber surfaces, it is not expected to significantly and directly modify the through-thickness capillary properties of the PCL membrane, which are important in filtering applications. While not verified by experimental tests in our study, this expectation is supported by models for disordered filtering media found in the literature, such as the work based on fractal theory reported in [25], linking the overall capillary flow to the through-thickness properties and, thus, suggesting that the nanoscale surface modification in our PCL-Ag scaffolds should not be detrimental in that regard.

The presence of silver on the surface of the PCL fibers was also confirmed by EDS mapping, as visible in Figure 4. Ag signal coming from the fiber bundle (Figure 4a) was detected in the EDS spectra (Figure 4b), and it was found to be distributed along the surface of the single fibers, as shown in Figure 4c–e.

SEM inspection of Ag deposited on reference silicon substrates, coated together with the PCL disks during the same process, was used to estimate the Ag content and demonstrate the formation of an incoherent nanoscale coating with the chosen sputtering parameters, that is both functional and "delicate" enough for polymer applications from a technological perspective. Low magnification micrograph in Figure 5a shows a uniform distribution of Ag particles over the substrate surface. Looking to the higher magnification (Figure 5b), the fine morphology appears constituted by islands formed as a results of silver atoms and ions coalescing on the silicon substrate during the deposition process, as discussed in previous work [19].

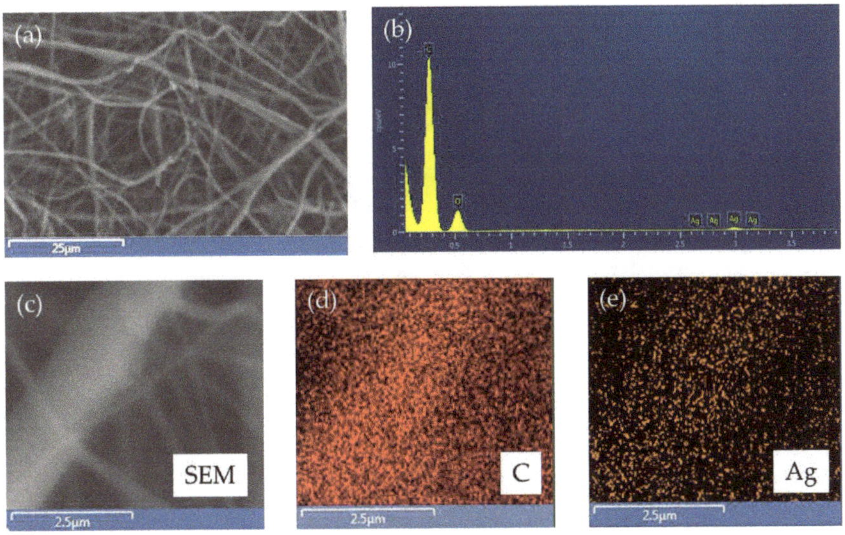

Figure 4. EDS mapping confirming presence of Ag on the Ag-coated PCL samples obtained at 5 kV operating voltage. (**a**,**b**) SEM and related EDS spectrum taken on a fiber bundle region; (**c**–**e**) SEM and related EDS mapping taken on a single fiber.

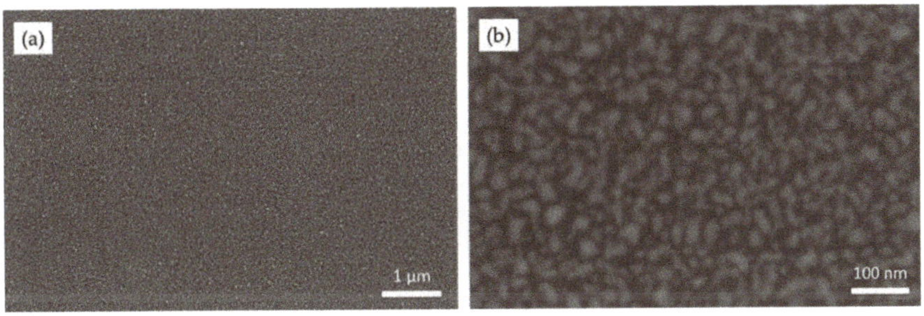

Figure 5. SEM images at (**a**) low and (**b**) high magnification of silver coating deposited on reference silicon substrate.

Measuring the substrate coverage α from high magnification SEM images, it was possible to estimate the Ag load per unit surface (concentration C in μg/cm^2) deposited on the samples, as:

$$C = \rho \times \alpha \times t, \quad (3)$$

where ρ is the silver density and t is the Ag coating thickness. By assuming the theoretical Ag density ρ = 10.49 g/cm^3, the coating thickness t = 6 nm as measured by atomic force microscopy [19], and α = 0.7 (being the effective coated surface around 70% as per SEM inspection), the specific density of the deposited Ag coating in our samples was about 4.4 μg/cm^2, corresponding to a remarkably low Ag load.

3.2. Wettability Analyses

The wettability of the blank and Ag-functionalized electrospun samples (PCL and PCL-Ag, respectively) was determined by contact angle measurements with water and diiodomethane. The photographs of representative measurements are shown in Figure 6, while the water contact angles (WCA) are presented in Table 2.

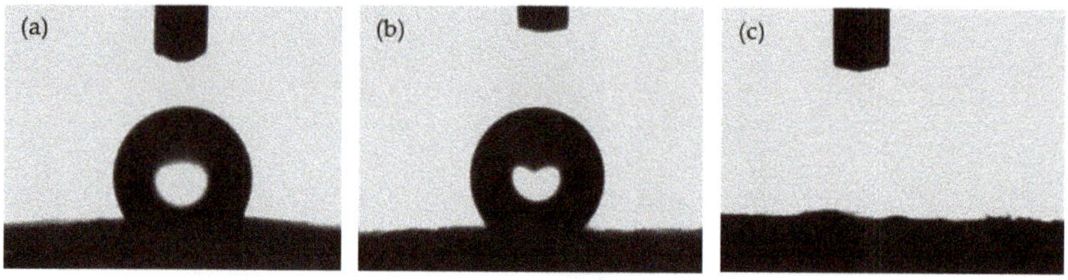

Figure 6. Photographs of water droplet on PCL surface before (**a**) and after Ag deposition (**b**). Photograph of diiodomethane droplet on blank PCL surface (**c**), as representative of both PCL and PCL-Ag surfaces.

Table 2. Water contact angles (mean ± standard deviation) of PCL and PCL-Ag samples.

Sample	WCA (°)
PCL	128.7 ± 4.7
PCL-Ag	134.3 ± 2.6

The pure PCL fibrous mats showed a water contact angle around 129°, indicating the hydrophobic nature of the studied PCL scaffolds, as expected due to the presence of CH_2 groups in the backbone of PCL chains, that is the main reason for its hydrophobic character [26]. After the Ag NPs deposition on the surface of the PCL fibers, a slight increase in the contact angle was observed, resulting in a measured WCA value around 134°. The WCA was measured after 10, 60 and 120 s and no variations of the drop shape were observed. The apolar behavior of PCL and PCL-Ag was confirmed by the full permeability displayed when diiodomethane was used to perform the test. In this case, the fluid was fully absorbed by both the uncoated and Ag-coated PCL fibers (Figure 6c).

It is worth noting that the slight increase of WCA after Ag addition by means of physical deposition techniques is consistent with results previously reported on polylactide (PLA) substrates [19]. This is particularly interesting when compared with Ag modification of PCL scaffolds by chemical-based routes, where the presence of silver turned the PCL wettability from hydrophobic to hydrophilic [11,14]. Thus, physical deposition of Ag NPs onto PCL fibers by the sputtering technique represents a successful method to retain the hydrophobic character in the final composites.

3.3. Antimicrobial Tests

The antimicrobial activity of the uncoated and Ag-coated PCL samples was tested for Gram-negative bacteria *E. Coli*. The samples were incubated with the microbial population, and the survival rate was determined at 2 and 4 h, using the count plate method and calculation of CFUs to determine the cytocitic action of the samples. Photographs of the Petri dishes obtained from control culture (*E. coli* in the absence of any mat) and from cultures with blank PCL and PCL-Ag mats at different serial dilutions and different sampled times are reported in Figure 7. Data obtained from the CFU calculations are listed in Table 3 and plotted in the top right panel of Figure 7.

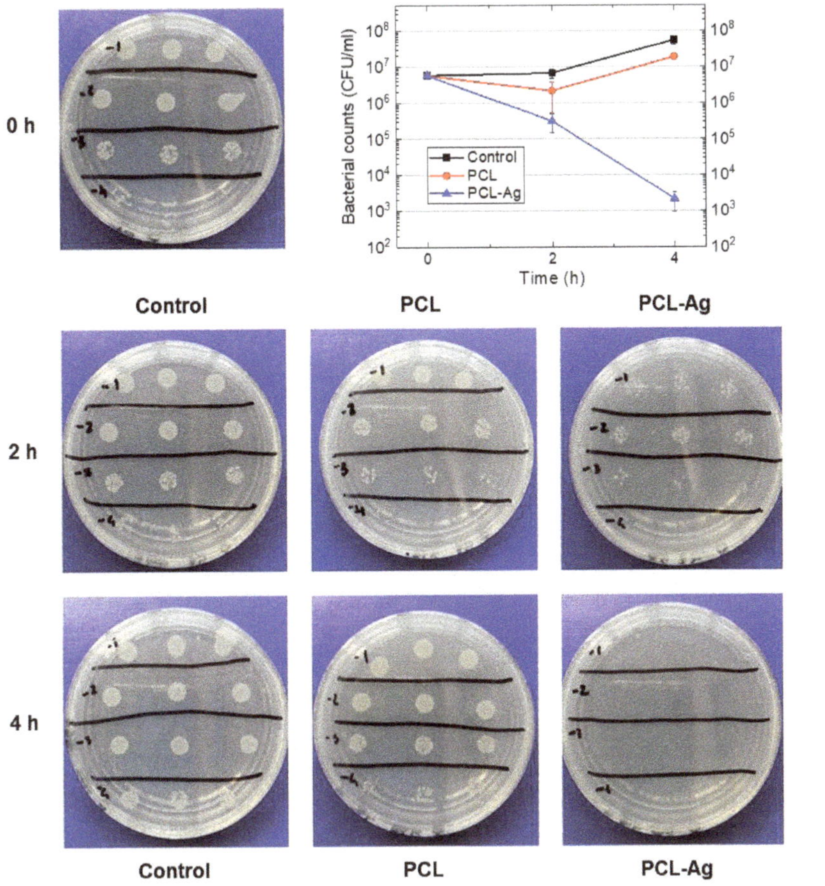

Figure 7. Photographs of Petri dishes obtained from *E. coli* culture without any mat (Control) and cultures with blank PCL and PCL-Ag mats at different serial dilutions from 10^{-1} to 10^{-4} (writing on the covers) and different sampled times (0, 2 and 4 h). Top right panel: graph of data obtained from the CFU calculations.

Table 3. Bacterial population (CFU/mL) measured for control, uncoated sample (PCL) and coated sample (PCL-Ag) at different times. Data are reported as mean ± standard deviations from two independent experiments, each in triplicate.

	Bacterial Population (CFU/mL)		
Sample	**0 h**	**2 h**	**4 h**
Control	$(5.8 \pm 1.1) \times 10^6$	$(6.6 \pm 1.9) \times 10^6$	$(5.5 \pm 1.4) \times 10^7$
PCL	-	$(2.1 \pm 1.6) \times 10^6$	$(1.9 \pm 0.1) \times 10^7$
PCL-Ag	-	$(3.1 \pm 1.6) \times 10^5$	$(2.1 \pm 1.2) \times 10^3$

As expected, *E. Coli* culture in the absence of any mat, used as a control, kept growing up to 4 h, passing from the initial concentration of 5.8×10^6 to 5.5×10^7 CFU/mL at 4 h. It was observed that the blank PCL sample induced a slight reduction in bacterial growth, leading to a 3-fold decrease of bacterial population with respect to the control. This weak antimicrobial action remained constant in the considered time range, indicating that the as-grown bare PCL mat represented a slightly unfavorable environment for bacterial growth, probably due to hindering in bacterial cells reproduction and to mechanical detrimental effects (e.g., cell deformation and wall abrasion caused by the interactions between the bacterial cells and the material surface). Interestingly, PCL-Ag sample showed good antimicrobial action already in the first 2 h of treatment, by reducing *E. coli* cellular density of almost 1 Log (about 7 times) compared to the blank PCL sample and more than 1 Log (about 20 times) compared to the control population. After 4 h the inhibitory effect of PCL-Ag on bacterial grown was excellent, leading to a decrease of almost 4 Log compared to blank PCL and almost 4.5 Log compared to the control. Such strong antibacterial activity was reached in few hours and by using a very lower amount of silver in comparison to previous findings [21] (nominal thickness of 6 nm vs. 50–275 nm). This can be a consequence of the Ag nanoparticle deposition achieved on the PCL fibers in our samples, instead of a uniform film coverage of the fibers, since nanostructuration leads to increase the effective surface of antibacterial agent exposed to the environment and the consequent antimicrobial efficacy [9]. Moreover, it is worth noting that the antibacterial effect in our PCL-Ag scaffolds resulted faster and higher with respect to composites produced starting from a mixed blend [11,13,27], despite the Ag content ratio in our samples (approximately 0.1 wt.% with respect to PCL) is comparable or even well below the one used in previous works, as expected due to the presence of the Ag nanoparticles directly on the surfaces of the PCL fibers in our method rather than incorporated within the fibers. More importantly, it must be underlined that, since we are dealing with Ag coating the fiber surfaces, the most correct parameter to be taken into account is essentially the Ag surface load in our PCL-Ag mats, that is as low as ~4 $\mu g/cm^2$, being able to confer the very effective antimicrobial action here demonstrated.

4. Conclusions

In this work we proposed the preparation of antibacterial mats through deposition of Ag NPs by sputtering silver onto the surface of electrospun PCL scaffolds. SEM analyses demonstrated that the Ag deposition process did not alter the overall morphology and texture of the original PCL mats characterized by randomly oriented fibers. The presence of Ag nanoparticles, with average size around 20 nm, decorating the surface of the PCL fibers was evident in SEM images and confirmed by EDS mapping. The wettability tests revealed that the hydrophobic character of the polymer was preserved in the final PCL-Ag composite. The Ag-coated PCL mats exhibited high antibacterial effect against gram-negative *E. coli* bacterial strain. This strong antimicrobial action obtained despite the low Ag content is ascribable to the increased effective surface of the antibacterial agent, thanks to the nanostructuration of silver in form of nanoparticles directly deposited on the surfaces of the PCL fibers.

On the basis of the PCL polymer properties, of the peculiar characteristics of the electrospun fiber scaffolds and of the antimicrobial activity provided by means of this straightforward method, the properties of the PCL-Ag systems developed and the proposed method can find promising exploitation in different applications, such as biomedical devices, medical PPE, food packaging, bioremediation and antifouling coatings in filtration.

The results here presented demonstrate that PCL can be successfully treated through an easy, flexible and effective method for the production of Ag-functionalized PCL mats endowed with interesting antimicrobial properties, as advocated in prior studies [22,23]. In order to confirm the practical usage of the materials here developed for the proposed applications, further analyses are expected to evaluate other important aspects, such as particle release phenomena, cytocompatibility with human cells, and coating stability.

Author Contributions: Conceptualization, D.V. and A.R.; methodology, D.V., L.T., R.V., G.G. and A.R.; investigation, D.V., L.T., R.V., G.G. and A.R.; writing—original draft preparation, D.V., L.T., R.V. and A.R.; writing—review and editing, D.V., L.T., R.V. and A.R. All authors have read and agreed to the published version of the manuscript.

Funding: This research received no external funding.

Institutional Review Board Statement: Not applicable.

Informed Consent Statement: Not applicable.

Data Availability Statement: Data available on request from the corresponding authors.

Conflicts of Interest: The authors declare the following competing financial interest(s): The authors A.R. and G.G. are affiliated with the company Nanofaber srl. This study is not a funded or sponsored research work.

References

1. Shin, Y.M.; Hohman, M.M.; Brenner, M.P.; Rutledge, G.C. Electrospinning: A whipping fluid jet generates submicron polymer fibers. *Appl. Phys. Lett.* **2001**, *78*, 1149–1151. [CrossRef]
2. Teo, W.E.; Ramakrishna, S. A review on electrospinning design and nanofibre assemblies. *Nanotechnology* **2006**, *17*, R89–R106. [CrossRef]
3. Mirjalili, M.; Zohoori, S. Review for application of electrospinning and electrospun nanofibers technology in textile industry. *J. Nanostruct. Chem.* **2016**, *6*, 207–213. [CrossRef]
4. Nitti, P.; Gallo, N.; Palazzo, B.; Sannino, A.; Polini, A.; Verri, T.; Barca, A.; Gervaso, F. Effect of L-Arginine treatment on the in vitro stability of electrospun aligned chitosan nanofiber mats. *Polym. Test.* **2020**, *91*, 106758. [CrossRef]
5. Haider, A.; Haider, S.; Kang, I.-K. A comprehensive review summarizing the effect of electrospinning parameters and potential applications of nanofibers in biomedical and biotechnology. *Arab. J. Chem.* **2018**, *11*, 1165–1188. [CrossRef]
6. Woodruff, M.A.; Hutmacher, D.W. The return of a forgotten polymer—Polycaprolactone in the 21st century. *Prog. Polym. Sci.* **2010**, *35*, 1217–1256. [CrossRef]
7. Mohamed, R.M.; Yusoh, K. A Review on the Recent Research of Polycaprolactone (PCL). *Adv. Mater. Res.* **2015**, *1134*, 249–255. [CrossRef]
8. Guarino, V.; Gentile, G.; Sorrentino, L.; Ambrosio, L. Polycaprolactone: Synthesis, Properties, and Applications. In *Encyclopedia of Polymer Science and Technology*; John Wiley and Sons, Inc.: Hoboken, NJ, USA, 2017; pp. 1–36.
9. Khezerlou, A.; Alizadeh-Sani, M.; Azizi-Lalabadi, M.; Ehsani, A. Nanoparticles and their antimicrobial properties against pathogens including bacteria, fungi, parasites and viruses. *Microb. Pathog.* **2018**, *123*, 505–526. [CrossRef]
10. Koutsoumpis, S.; Poulakis, A.; Klonos, P.; Kripotou, S.; Tsanaktsis, V.; Bikiaris, D.N.; Kyritsis, A.; Pissis, P. Structure, thermal transitions and polymer dynamics in nanocomposites based on poly(ε-caprolactone) and nano-inclusions of 1-3D geometry. *Thermochim. Acta* **2018**, *666*, 229. [CrossRef]
11. Tran, P.A.; Hocking, D.M.; O'Connor, A.J. In situ formation of antimicrobial silver nanoparticles and the impregnation of hydrophobic polycaprolactone matrix for antimicrobial medical device applications. *Mater. Sci. Eng. C* **2015**, *47*, 63. [CrossRef]
12. Sumitha, M.S.; Shalumon, K.T.; Sreeja, V.N.; Jayakumar, R.; Nair, S.V.; Menon, D. Biocompatible and antibacterial nanofibrous poly(ε-caprolactone)- nanosilver composite scaffolds for tissue engineering applications. *J. Macromol. Sci. Part A Pure Appl. Chem.* **2012**, *49*, 131. [CrossRef]
13. Pazos-Ortiz, E.; Roque-Ruiz, J.H.; Hinojos-Márquez, E.A.; López-Esparza, J.; Donohué-Cornejo, A.; Cuevas-González, J.C.; Espinosa-Cristóbal, L.F.; Reyes-López, S.Y. Dose-Dependent Antimicrobial Activity of Silver Nanoparticles on Polycaprolactone Fibers against Gram-Positive and Gram-Negative Bacteria. *J. Nanomater.* **2017**, *2017*, 4752314. [CrossRef]
14. Thomas, R.; Soumya, K.R.; Mathew, J.; Radhakrishnan, E.K. Electrospun Polycaprolactone Membrane Incorporated with Biosynthesized Silver Nanoparticles as Effective Wound Dressing Material. *Appl. Biochem. Biotechnol.* **2015**, *176*, 2213. [CrossRef]
15. Lim, M.M.; Sultana, N. In vitro cytotoxicity and antibacterial activity of silver-coated electrospun polycaprolactone/gelatine nanofibrous scaffolds. *3 Biotech* **2016**, *6*, 211. [CrossRef]
16. Valerini, D.; Tammaro, L.; Villani, F.; Rizzo, A.; Caputo, I.; Paolella, G.; Vigliotta, G. Antibacterial Al-doped ZnO coatings on PLA films. *J. Mater. Sci.* **2020**, *55*, 4830–4847. [CrossRef]
17. Valerini, D.; Tammaro, L.; Di Benedetto, F.; Vigliotta, G.; Capodieci, L.; Terzi, R.; Rizzo, A. Aluminum-doped zinc oxide coatings on polylactic acid films for antimicrobial food packaging. *Thin Solid Films* **2018**, *645*, 187–192. [CrossRef]
18. Benetti, G.; Cavaliere, E.; Banfi, F.; Gavioli, L. Antimicrobial nanostructured coatings: A gas phase deposition and magnetron sputtering perspective. *Materials* **2020**, *13*, 784. [CrossRef]
19. Valerini, D.; Tammaro, L.; Vigliotta, G.; Picariello, E.; Banfi, F.; Cavaliere, E.; Ciambriello, L.; Gavioli, L. Ag functionalization of Al-doped ZnO nanostructured coatings on PLA substrate for antibacterial applications. *Coatings* **2020**, *10*, 1238. [CrossRef]
20. Memic, A.; Aldhahri, M.; Tamayol, A.; Mostafalu, P.; Abdel-Wahab, M.S.; Samandari, M.; Moghaddam, K.M.; Annabi, N.; Bencherif, S.A.; Khademhosseini, A. Nanofibrous silver-coated polymeric scaffolds with tunable electrical properties. *Nanomaterials* **2017**, *7*, 63. [CrossRef]
21. Kalakonda, P.; Aldhahri, M.A.; Abdel-Wahab, M.S.; Tamayol, A.; Moghaddam, K.M.; Ben Rached, F.; Pain, A.; Khademhosseini, A.; Memic, A.; Chaieb, S. Microfibrous silver-coated polymeric scaffolds with tunable mechanical properties. *RSC Adv.* **2017**, *7*, 34331. [CrossRef]

22. Rumbo, C.; Tamayo-Ramos, J.A.; Caso, M.F.; Rinaldi, A.; Romero-Santacreu, L.; Quesada, R.; Cuesta-López, S. Colonization of Electrospun Polycaprolactone Fibers by Relevant Pathogenic Bacterial Strains. *ACS Appl. Mater. Interfaces* **2018**, *10*, 11467–11473. [CrossRef] [PubMed]
23. Tamayo-Ramos, J.A.; Rumbo, C.; Caso, M.F.; Rinaldi, A.; Garroni, S.; Notargiacomo, A.; Romero-Santacreu, L.; Cuesta-López, S. Analysis of Polycaprolactone Microfibers as Biofilm Carriers for Biotechnologically Relevant Bacteria. *ACS Appl. Mater. Interfaces* **2018**, *10*, 32773–32781. [CrossRef]
24. Pagnotta, G.; Graziani, G.; Baldini, N.; Maso, A.; Focarete, M.L.; Berni, M.; Biscarini, F.; Bianchi, M.; Gualandi, C. Nanodecoration of electrospun polymeric fibers with nanostructured silver coatings by ionized jet deposition for antibacterial tissues. *Mater. Sci. Eng. C* **2020**, *113*, 110998. [CrossRef]
25. Xiao, B.; Huang, Q.; Chen, H.; Chen, X.; Long, G. A fractal model for capillary flow through a single tortuous capillary with roughened surfaces in fibrous porous media. *Fractals* **2021**, *29*, 2150017. [CrossRef]
26. Zhang, J.; Qiu, Z. Morphology, crystallization behavior, and dynamic mechanical properties of biodegradable poly(ε-caprolactone)/thermally reduced graphene nanocomposites. *Ind. Eng. Chem. Res.* **2011**, *50*, 13885–13891. [CrossRef]
27. Gao, Y.; Hassanbhai, A.M.; Lim, J.; Wang, L.; Xu, C. Fabrication of a silver octahedral nanoparticle-containing polycaprolactone nanocomposite for antibacterial bone scaffolds. *RSC Adv.* **2017**, *7*, 10051. [CrossRef]

coatings

MDPI

Article

Bodipy-Loaded Micelles Based on Polylactide as Surface Coating for Photodynamic Control of *Staphylococcus aureus*

Enrico Caruso [1], Viviana Teresa Orlandi [1], Miryam Chiara Malacarne [1], Eleonora Martegani [1], Chiara Scanferla [1], Daniela Pappalardo [2], Giovanni Vigliotta [3] and Lorella Izzo [1,*]

1 Dipartimento di Biotecnologie e Scienze della Vita, Università degli Studi dell'Insubria, 21100 Varese, Italy; enrico.caruso@uninsubria.it (E.C.); viviana.orlandi@uninsubria.it (V.T.O.); mc.malacarne@uninsubria.it (M.C.M.); e.martegani@uninsubria.it (E.M.); cscanferla@uninsubria.it (C.S.)

2 Dipartimento di Scienze e Tecnologie, Università degli Studi del Sannio, 82100 Benevento, Italy; pappal@unisannio.it

3 Dipartimento di Chimica e Biologia, Università degli Studi di Salerno, 84084 Fisciano, Italy; gvigliotta@unisa.it

* Correspondence: lorella.izzo@uninsubria.it

Abstract: Decontaminating coating systems (DCSs) represent a challenge against pathogenic bacteria that may colonize hospital surfaces, causing several important infections. In this respect, surface coatings comprising photosensitizers (PSs) are promising but still controversial for several limitations. PSs act through a mechanism of antimicrobial photodynamic inactivation (aPDI) due to formation of reactive oxygen species (ROS) after light irradiation. However, ROS are partially deactivated during their diffusion through a coating matrix; moreover, coatings should allow oxygen penetration that in contact with the activated PS would generate $^1O_2$, an active specie against bacteria. In the attempt to circumvent such constraints, we report a spray DCS made of micelles loaded with a PS belonging to the BODIPY family (2,6-diiodo-1,3,5,7-tetramethyl-8-(2,6-dichlorophenyl)-4,4′-difluoroboradiazaindacene) that is released in a controlled manner and then activated outside the coating. For this aim, we synthesized several amphiphilic copolymers (mPEG–(PLA)$_n$), which form micelles, and established the most stable supramolecular system in terms of critical micelle concentration (CMC) and ΔG_f values. We found that micelles obtained from mPEG–(PLLA)$_2$ were the most thermodynamically stable and able to release BODIPY in a relatively short period of time (about 80% in 6 h). Interestingly, the BODIPY released showed excellent activity against *Staphylococcus aureus* even at micromolar concentrations.

Keywords: polylactide; photodynamic; supramolecular systems; micelles; drug delivery; copolymers; ring opening polymerization; aPDT; BODIPY

Citation: Caruso, E.; Orlandi, V.T.; Malacarne, M.C.; Martegani, E.; Scanferla, C.; Pappalardo, D.; Vigliotta, G.; Izzo, L. Bodipy-Loaded Micelles Based on Polylactide as Surface Coating for Photodynamic Control of *Staphylococcus aureus*. *Coatings* **2021**, *11*, 223. https://doi.org/10.3390/coatings11020223

Academic Editor: Sami Rtimi

Received: 29 December 2020
Accepted: 8 February 2021
Published: 13 February 2021

1. Introduction

One of the route of pathogen dissemination is via contaminated surfaces, as most bacteria can survive for a long time even on the surface of objects. Nosocomial infections generated by contaminated surfaces are a great concern because common disinfectants used in routine cleansing are generally based on quaternary ammonium compounds (QACs), halogen releasing agents, and phenolics, in some cases resulting in ineffective killing of pathogenic bacteria, as their activity depends on several factors, including the surface contact period, pH, temperature, and amount and nature of the microorganism [1].

In the past two decades, to obtain decontamination systems, different antimicrobial substances have been incorporated in the bulk or in the coating of biomaterials. They can be added during the phase of production, a posteriori absorbed in biomaterials, or covalently bound to functionalized polymeric coatings [2].

The storage of anti-pathogens in bulk materials is one of the approaches used for the purpose of gradually releasing biocide, which provides sustained release that kills

pathogens. However, an excessive use of antibiotics leads to the development of antibiotic-resistant pathogens and could causes environmental contamination [3].

Among biocides that can be encapsulated as alternatives to antibiotics, such as phytocomplex, natural, and synthetic peptides [4–6], photosensitizers (PSs) represent a very promising alternative strategy based on the antimicrobial photodynamic inactivation (aPDI) of microorganism. PSs are molecules that, once irradiated with light of a specific wavelength, can reach a relatively long-lived triplet state able to transfer energy to molecular oxygen. The process leads to the formation of the reactive singlet oxygen ($^1O_2$) or, more in general, to reactive oxygen species (ROS) that together induce oxidation reactions resulting in the death of bacteria. Recently, materials containing PSs have been synthesized, characterized, and tested as coatings with antimicrobial efficiency for decontaminating surfaces. Although this strategy allows for the control of multidrug-resistant strains and the inhibition of biofilm formation [7], several requirements limit its application.

To deliver PSs, Hamblin [8] recently proposed some types of nano-drug systems, such as liposomes and micelles. For example, the photoinactivation of *Staphylococcus aureus* by chlorophyll derivatives was increased by the use of liposomes compared to micelles, suggesting that this delivery system is promising in aPDI applications [9]. Furthermore, light-based techniques are believed to be a potent method to counteract microbial biofilms. Thus, the delivery of PSs represents an important goal to reach. To this aim, the PS chlorin e6 (Ce6) was grafted to α-cyclodextrin (α-CD)/polyethylene glycol (PEG). The supramolecular micellar assembly of Ce6 showed enhanced inhibition of biofilm formation of methicillin-resistant *Staphylococcus aureus* (MRSA) [10]. In clinical applications, new delivery systems could overcome undesired damage to normal cells, ascribable to the nonselective nature of photodynamic therapy (PDT). In addition, in decontaminating coating systems based, for example, on polymer matrixes, PSs may be located in polymer layers or polymers fibers. The antimicrobial activity is generated by the PDI process following ROS diffusion through the matrix.

McCoy et al. [11] reported a two-layer system based on PSs incorporated into high density polyethylene (HDPE) through hot-melt extrusion. One layer consisted of a PS contained in HDPE, while the second was made with HDPE alone. Systems containing different PSs were able to reduce the adherence of viable bacteria by up to 3.62 Log colony forming units (CFU) per cm^2 of surface for methicillin-resistant *Staphylococcus aureus* (MRSA) and by up to 1.51 Log CFU/cm^2 for *Escherichia coli*. However, they found that PS chemical compatibility with HDPE was essential to provide complete miscibility of PS with the polymer and consequently homogeneity in PS distribution into the matrix; additionally, with some of the PSs used, the surface of materials showed roughness and irregularity that promoted adhesion of bacteria.

Recently, electrospinning-based technology allowed PS-doped electro-spun materials to be developed, and consequently several systems containing PSs have been proposed as antimicrobial coatings. In this case, photodynamic antimicrobial activity was compromised by the presence of photodynamic molecules into nanofibers, considering that ROS possess short lifetime (less than 3.5 μs) and short diffusion length (about 10–100 nm) [12–17].

In this framework, Felgenträger et al. [18] developed a polymeric surface coating consisting of a derivative of meso-tetraphenyl porphyrin (TPP) immobilized onto polyurethane (PU) after being sprayed and polymerized as a thin layer onto a surface of poly-methylmethacrylate (PMMA). PU is gas permeable and thus was chosen to guarantee that enough oxygen reached the PS into the coating to produce $^1O_2$. Even if the diffusion of ROS was proved by the reduction of bacteria on the surface, the $^1O_2$ luminescence signal exhibited multi-exponential decay, which the authors attributed to a deactivation in the surroundings, such as PU, PMMA, or air.

To overcome limitations of polymer coatings containing PSs, we report here a spray-coating system based on degradable polymer micelles made of poly(ethylene glycol) monomethylether-*co*-polylactide branched copolymers of the general structure mPEG–(PLA)$_n$ (Mn_{mPEG} = 5 kDa, n = 1, 2, or 4), releasing a PS. The same copolymers were previously obtained in both micro- and nanometric formulations and previously described also as drug delivery systems [19–21]. The photoactive molecule was chosen among the 4,4-difluoro-4-bora-3a,4a-diaza-s-indacene (BODIPY) dyes family, which are characterized to be strongly UV-absorbing and to have high quantum efficiencies of fluorescence. They are also relatively insensitive to the polarity and pH of the environment and are reasonably stable in physiological conditions. BODIPY-based dyes are generally considered an interesting class of compounds for antimicrobial photodynamic therapy (aPDT), as they maintain photochemical properties, even when modified, by introducing substituent groups in some positions of the molecule [22–24]. In the present study, micelle coating surfaces worked as micro-reservoirs of BODIPY, which was released on the surface in a controlled manner.

Overall, surface decontamination with the micelle coating was thought as a process consisting of the release of BODIPY in a relatively short period of time during which the PS is continuously activated by irradiation. The aim is to realize a decontamination treatment through a "spray-coating" capable of removing bacteria completely and deeply from surfaces, avoiding ROS diffusion through the coating matrix and their partial deactivation.

2. Materials and Methods

2.1. Materials

All manipulations involving air-sensitive compounds were carried out under nitrogen atmosphere using Schlenk or dry-box techniques. Poly(ethylene glycol) monomethyl ether (mPEG, M_n = 5000 Da) was purchased from Aldrich and dried in vacuo over phosphorus pentoxide for 72 h at 25 °C prior to use (Aldrich). L- and D,L-lactide, purchased from Aldrich, were crystallized from dry toluene, then dried in vacuo with phosphorus pentoxide for 72 h at 25 °C. Toluene, purchased from Aldrich, was dried over sodium and distilled before use. $Al(CH_3)_3$, phosphorus pentoxide, and pyrene were supplied from Aldrich and used as received. Dichloromethane (DCM) and dimethyl sulfoxide (DMSO) were purchased from Carlo Erba and used without purification. Dialysis was performed with Orange Scientific membrane and OrDial D35-MWCO 3500 regenerated cellulose dialysis tubing. The compound 2,6-diiodo-1,3,5,7-tetramethyl-8-(2,6-dichlorophenyl)-4,4′-difluoroboradiazaindacene was synthetized as previously reported [25].

2.2. Synthesis of mPEG–(PLA)$_n$ Copolymers

The copolymers used in this work were synthesized according to the procedures reported in the literature [19] by using a PEG/LA ratio (*w*/*w*) of 1/2.

A typical procedure is herein described for the copolymer of entry 1 and 2 (Table 1). A magnetically stirred reactor vessel (50 mL) was charged sequentially with a solution of mPEG (M_n = 5 kDa, 0.500 g, 0.1 mmol) and $AlMe_3$ (25 mg; 0.35 mmol) in toluene (10 mL). The mixture was stirred for 1 h, and then L- or D,L-lactide (1.00 g, 6.94 mmol) was added and the mixture was magnetically stirred at 70 °C for 48 h. Conversions were monitored by integration of the monomer vs. polymer methine resonances in the ^{1}H NMR spectrum of crude product (in $CDCl_3$). At complete monomer conversion, the mixture was poured into hexane (200 mL), and the precipitated polymer was recovered by filtration, washed with methanol, and dried at 40 °C in a vacuum oven. The copolymer was characterized by ^{1}H NMR spectroscopy.

^{1}H NMR ($CDCl_3$, 25 °C) δ = 1.54 (m, –$CHCH_3$–), 3.64 (s, –CH_2–), 5.15 (m, –$CHCH_3$–).

The molar percentage of lactide in each copolymer was evaluated by the relative intensity of signals at δ: 5.15 (–CH– of polylactide, A_{CH}) and 3.64 (–CH_2– of m-PEG, A_{CH2}).

Molecular weight was evaluated by ^{1}H NMR using the following equation:

$$Mn = (Mn_{PEG}\ (MW_{lactide})(2A_{CH}))/((A_{CH2})\ (MW_{ethyleneoxide})) + Mn_{PEG} \quad (1)$$

2.3. NMR Measurements

The ^{1}H-NMR spectra were recorded on a Bruker Avance 400 spectrometer (^{1}H, 400.00 MHz) at 25 °C using tetramethyl silane (TMS) as an internal reference (Aldrich).

Samples were prepared by introducing 20 mg of copolymer and 0.5 mL of $CDCl_3$ into an NMR-tube (5 mm outer diameter).

2.4. CMC Measurements

The critical micelle concentration (CMC) was determined using pyrene as a fluorescence probe. Samples for fluorescence spectroscopy were prepared by diluting the micelle solutions to 10 different concentrations (range 1.0×10^{-4} mol/L–1.0×10^{-7} mol/L). Each sample was then obtained by dropping a pyrene solution (5.0×10^{-6} mol/L in acetone) into an empty vial, adding one of the copolymer solutions previously prepared, and evaporating the acetone by gentle heating. The volume of the needed copolymer solution was calculated to have a final pyrene concentration in water of 6.0×10^{-7} mol/L, which is slightly below the pyrene saturation concentration at 22 °C. Fluorescence spectra were recorded using a Varian luminescence spectrometer at an excitation wavelength of 335 nm at 22 °C. The intensities of the bands I_1 at 372 nm and I_3 at 383 nm were evaluated and their ratios plotted vs. the copolymer concentration.

2.5. Preparation of BODIPY Loaded Micelles and Study of Release

A total of 800 μL of a 1 mg/mL solution of BODIPY in DCM and 1 mL of a solution containing the copolymer (solution prepared with 20 mg of polymer dissolved in 1 mL of DCM) were introduced in a dialysis membrane and dialyzed against PBS (pH 7.4) for 24 h.

In order to eliminate the residue of BODIPY, at the end of the dialysis, the sample was recovered from the membrane and centrifuged at 1000 rpm at 4 °C for 10 min. The supernatant was collected, and then 300 μL of a 2:1 DMSO:PBS solution was added to 300 μL of supernatant to determine the concentration of BODIPY loaded into the micelles.

The BODIPY loading was evaluated using the following equation:

$$\text{BODIPY loading } (\%) = \left[\frac{\text{mol of BODIPY loaded}}{\text{mol of micelles}}\right] \times 100 \quad (2)$$

To evaluate the amount of BODIPY released over time, at fixed times the solution containing the loaded micelles was centrifuged at 500 rpm at 5 °C for 5 min; the supernatant was removed while the precipitate was dissolved in 600 μL of 2:1 DMSO:PBS solution and analyzed by UV–vis spectroscopy at the λ = 548 nm of BODIPY maximum absorbance to determine the BODIPY concentration

2.6. Bacterial Strain and Growth Conditions

The photoinactivation assays were performed against the model pathogen *Staphylococcus aureus* ATCC6538P. *S. aureus* was grown overnight in tryptic soy broth (TSB) medium at 37 °C on an orbital shaker at 200 rpm. When necessary, a solid formulation (agar 15% *w*/*v*) was prepared.

2.7. Photoinactivation by Suspended Micelles

Upon overnight growth, *S. aureus* cultures were centrifuged (8000 rpm, 4 min), and the pellet was suspended in phosphate buffer saline (PBS, KH_2PO_4/K_2HPO_4 10 mM, pH 7.4) to reach a bacterial concentration of ~10^8 CFU/mL. BODIPY (1 or 0.1 μM) dissolved in DMSO or micelle encapsulated BODIPY (0.1 μM) was administered to cell suspensions. Samples were dark incubated for 10 min to favor the interaction between PSs and bacterial cells and then irradiated, at increasing times, under 520 nm light (2.4 mW/cm^2). At 1 h of irradiation, the final light dose was 8.7 J/cm^2, and at 6 h it was 52.1 J/cm^2. A panel of the following controls was kept under dark incubation: untreated cells, cells treated with DMSO (0.01% *v*/*v*) or with empty micelles, and cells treated with BODIPY (dissolved in DMSO or encapsulated). The effect of irradiation was also evaluated on untreated cells and cells treated with DMSO or empty micelles.

To determine the effect of dark incubation and irradiation on bacterial viability of control samples and PS-treated samples, cellular concentration was determined by viable count technique. Each sample was collected and 10-fold serially diluted in PBS. A volume of 10 μL of each diluted and undiluted sample was plated on TSB agar plates. After overnight incubation at 37 °C, the cellular concentration was calculated and expressed as colony forming units per milliliter (CFU/mL). Photoinactivation experiments were performed at least in triplicate.

2.8. Photoinactivation Induced by Micelle-Coated Glass

A volume of 200 μL of PS-loaded micelles or free PS (0.8 μM) was sprayed on coverslip glasses. A sample without any coating was also included. Coated glasses were dried under airflow and positioned in 35 mm diameter petri dishes. Each dish was filled with 1.5 mL of bacterial suspension prepared by diluting 300-fold with an overnight *S. aureus* ATCC 6538P culture in M9 minimal medium added with 10 mM glucose and 0.2% *w/v* casamino acids. After overnight incubation at 37 °C, coverslip glasses were irradiated under light at 520 nm for 2 h in a humid chamber. Upon irradiation, adherent cells were stained for 30 min with 2 μM fluorochrome 4,4-difluoro-1,3,5,7-tetramethyl-8-(2-methoxyphenyl)-4-bora-3a,4a-diaza-s-indacene [26]. Then, the coverslips were transferred on microscope glass slides for confocal microscopy analysis. The images were acquired through a 63X objective lens and a 488 nm laser.

3. Results and Discussion

Most micelle coatings reported in the literature consist of physical adsorption of micelles to the surface [27–29]; in a few cases, the chemical coating is derived by the reaction of side or terminal groups of block copolymers forming micelles with groups on surfaces [30–32]. The general advantage of using micelles consists of their supramolecular typical structure that allows for the holding and releasing of active molecules.

In this study, the coating comprised micelles physically adsorbed on the surface and loaded with BODIPY. Controlled release of BODIPY in a relatively short range of time, during which the free molecule is activated by irradiation, allows bacteria decontamination of the surface, as reported in Scheme 1, avoiding the drawbacks connected to the $^1O_2$ diffusion through a polymer matrix and its partial deactivation.

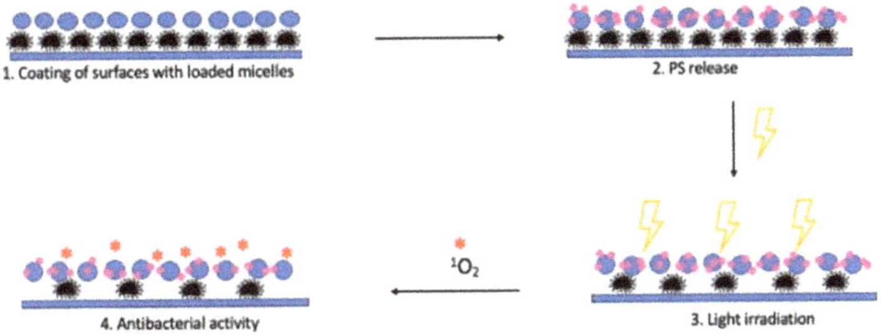

Scheme 1. Mechanism of action of the coating systems based on loaded micelles.

To this aim, we synthesized several linear and branched block copolymers of general structure AB_n (n = 1, 2, 4), where A is the poly(ethylene glycol)monomethylether (mPEG, 5 kDa) and B represents the poly(lactide) (PLA) blocks, either atactic poly D,L-lactide (PD,LLA) or stereoregular polyL-lactide (PLLA) (Figure 1). mPEG–$(PLA)_n$ linear and branched copolymers were synthesized following a procedure already reported by some of us [19] and with the rationale of obtaining a similar molecular weight for the hydrophobic and hydrophilic part of the copolymer to foster the formation of micelles by copolymer self-assembling in water. The idea of using micelles of mPEG–$(PLA)_n$ copolymers as surface releasing coatings is based on two consideration: the first one is about the degradability of copolymers, due to the presence of the polylactide blocks, that leads to a temporary coating; the second one is the presence of the mPEG block that preserves the surface from adsorption of proteins and adhesion of cells after its decontamination and until the following process of surface disinfection. The linear and branched structures were synthesized, since we already noted that the copolymer structure can strongly influence chemical–physical properties as well as the biological behavior of supramolecular aggregates, such as micelles or vesicles or both when the copolymer is neutral and charged [19–21,33–43].

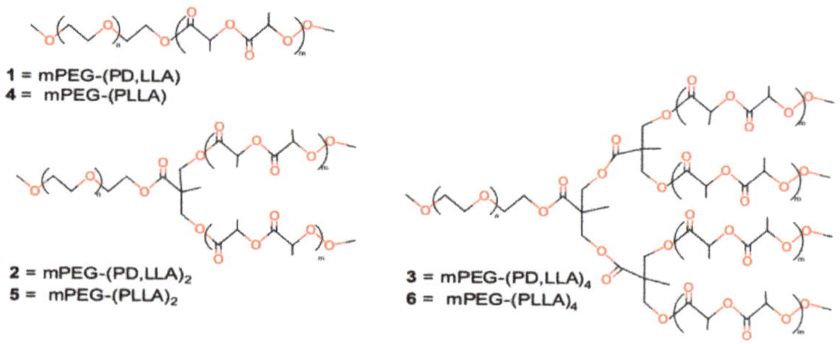

Figure 1. Copolymer structures. The numbers refer to the samples described in the relative entry of Table 1.

Table 1 shows data on microstructural features of the copolymers, the CMC of the corresponding micelles, and their ΔG formation. The last parameters are indicative of the thermodynamic stability of micelles and are fundamental for selecting copolymers that would maintain the supramolecular structure longer, allowing for the controlled release of PSs on the surface. The lower the CMC and ΔG values are, the higher the micelle stability is. It is well known that by controlling the crystallinity of the core it is possible to induce micelle stability. Data show that copolymers having the same general structure $A(B)_n$ possess lower CMC and ΔG when the polylactide chain is stereoregular (PLLA chains), which is noticeable when comparing CMC and ΔG values of entry 1 and 4, 2 and 5, and 3 and 6. Indeed, previous literature results showed that micelles formed with amorphous cores exhibited considerably higher CMCs than those with semicrystalline cores [44]. Among the stereoregular copolymers, mPEG–$(PLLA)_2$ and mPEG–$(PLLA)_4$ appeared the most appropriate for this study because of their lower CMC and ΔG. They are characterized by similar molecular weights but a different degree of branching and consequently different lengths of branches (17.4 and 13.4 monomer per arm for mPEG–$(PLLA)_2$ and mPEG–$(PLLA)_4$, respectively). It is worth noting that the two-arm branched, stereoregular copolymer mPEG–$(PLLA)_2$ shows the lowest CMC and ΔG, in accordance with what was previously found by some of us [20]. This experimental observation might be due to the presence of longer hydrophobic PLLA chains in the case of the two-arm branched copolymer compared to the four-arm branched ones. The longer PLLA arms may give rise to more inter-chain polymer entanglements, thus stabilizing the core and decreasing the CMC.

Table 1. Chemical structure, lactide content, and molecular weight (M_n) of copolymers. Critical micelle concentration (CMC) and ΔG formation of the corresponding micelles.

Entry	Copolymer [1]	LA unit/arm	M_n [2] (kDa)	CMC (M)	CMC (μg/mL)	ΔG [3] (kJ/mol)	BODIPY Loading [4] (%)
1	mPEG–(PD,LLA)	37.5	10.4	6.7×10^{-6}	70	−29.50	–
2	mPEG–$(PD,LLA)_2$	17.4	10.0	7.1×10^{-6}	71	−29.36	–
3	mPEG–$(PD,LLA)_4$	14.2	13.2	3.4×10^{-6}	39	−31.18	–
4	mPEG–(PLLA)	33.3	9.8	5.5×10^{-6}	54	−29.99	–
5	mPEG–$(PLLA)_2$	17.4	10.0	4.4×10^{-7}	4.4	−36.24	3.01
6	mPEG–$(PLLA)_4$	13.4	12.7	2.8×10^{-6}	36	−31.66	3.54

[1] mPEG molecular weight 5 kDa, [2] Evaluated by ^{1}H-NMR using Equation (1), [3] G = RTlnCMC, where CMC is expressed as mole fraction; T = 298 K, [4] Evaluated using Equation (2).

We loaded micelles obtained by self-assembling mPEG–(PLLA)$_2$ and mPEG–(PLLA)$_4$ in PBS (pH 7.4) with 2,6-diiodo-1,3,5,7-tetramethyl-8-(2,6-dichlorophenyl)-4,4′-difluoroboradiazaindacene (Figure 2), a PS belonging to the BODIPY family. This molecule showed a very high photodynamic activity against tumor cells, such us ovarian carcinoma cells (SKOV3), after irradiation with a green LED device [45], and thus it would be interesting to evaluate its antimicrobial activity against *Staphylococcus aureus* to assess whether neutral PSs, having an excellent antitumor efficacy, can also exert a strong antibacterial action. Our interest into antimicrobial activity against such bacteria derives from the fact that these pathogens are etiological agents of several nosocomial infections. Furthermore, the selection of strains resistant to a newer generation of antibiotics and the ability of *S. aureus* strains to form biofilm on inert surfaces and biological tissues makes their eradication more difficult [46].

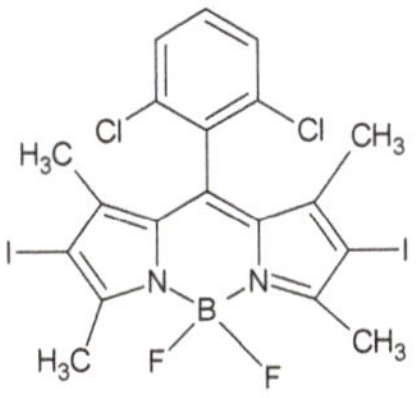

Figure 2. Structure of 2,6-diiodo-1,3,5,7-tetramethyl-8-(2,6-dichlorophenyl)-4,4′-difluoroboradiazaindacene (BODIPY) used as a photosensitizer (PS).

Table 1 shows data relative to the BODIPY loading for the two kind of copolymer forming micelles. The loading was quite similar for both micelles with a slightly higher amount of dye encapsulated in the more branched system. BODYPY release was studied in PBS solution at pH 7.4 and at room temperature (298 K) for 48 h, and it is reported in Figure 3. It is worth noting that both micellar systems could release the molecular dye in a controlled manner, with a massive but still contained release within the first half an hour, which was around 16% and 13% of the total content of BODIPY for mPEG–(PLLA)$_2$ and mPEG–(PLLA)$_4$ micelles, respectively. Interestingly, despite the lower amount of dye and higher stability, mPEG–(PLLA)$_2$ micelles were able to release almost 100% of BODIPY in the first 48 h, in contrast with the more branched ones that released only the 67% of total dye encapsulated in the same period of time. It is difficult to explain the different behavior of the two kinds of micelles in the release of BODIPY. One can suppose that the higher stability of mPEG–(PLLA)$_2$ micelles might be associated with the PLLA branch interactions in the core, which are stronger than those in the mPEG–(PLLA)$_4$. Thus, the strongest branch interactions may induce BODIPY molecules to accumulate preferentially in the outer part of the hydrophobic core, meaning close to the confined region between the hydrophobic region and the mPEG corona, during self-aggregation of mPEG–(PLLA)$_2$ copolymer.

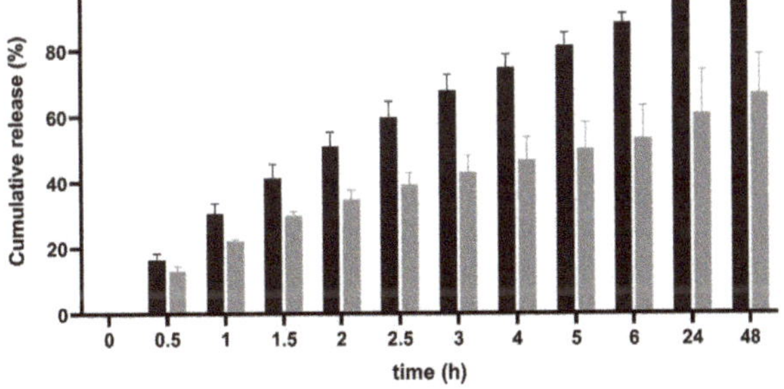

Figure 3. BODIPY release at pH 7.4 and 298 K from micelles made of mPEG–(PLLA)$_2$ (black) and mPEG–(PLLA)$_4$ (grey).

The different behavior in releasing BODIPY, as showed by the two systems, are both intriguing for a coating that requires a slow release of the antimicrobial PS. However, we decided to test the antimicrobial efficiency of the system based on the more stable micelles that release completely within 48 h.

To this aim, the efficiency of BODIPY in photoinactivation of the model strain *S. aureus* ATCC 6538P was tested, and data are reported in Figure 4. Under a low energy dose of light (8.7 J/cm^2) at 520 nm, BODIPY showed a dose-dependent killing effect. The lowest tested concentration (0.1 μM) caused a statistically significant reduction (p = 0.0361) of ~4 log unit, while the highest concentration (1 μM) caused the highest killing rate detectable in the chosen setup, namely ≤100 CFU/mL (p = 0.0023). All the included controls ruled out any intrinsic activity of BODIPY and of solvent used to solubilize PS (DMSO 0.01%). As expected, green light did not influence *S. aureus* viability. It is well known that neutral PSs belonging to several families, natural and synthetic, such as curcumin, hypericin, phthalocyanine, and fullerenes, are active against Gram-positive bacteria [47]. To the best of our knowledge, in this study, for the first time a neutral BODIPY was successfully tested as a PS against *S. aureus.* It is noteworthy to point out that the BODIPY was active at low micromolar concentration and was active only under irradiation. These features make the chosen BODIPY an interesting antimicrobial PS.

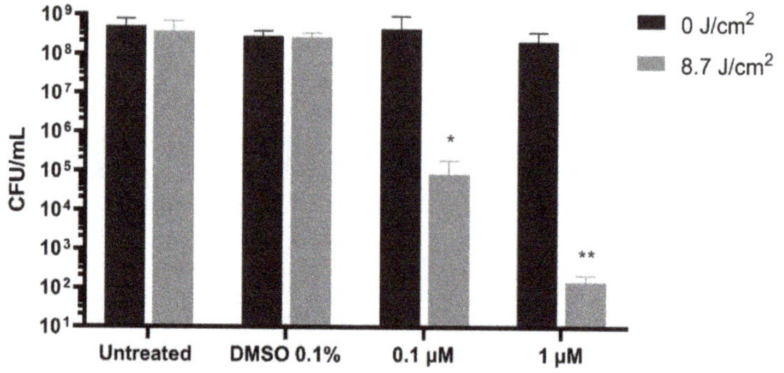

Figure 4. Photodynamic inactivation of *Staphylococcus aureus* by 0.1 and 1 μM of BODIPY dissolved in DMSO. Irradiation with light at 520 nm was performed for 1 h to reach a final dose of 8.7 J/cm^2. Untreated and DMSO (0.1% v/v)-treated samples were included as controls. Data of cell viability are expressed as CFU/mL ± standard deviations. Statistical analyses were performed by one-way ANOVA (* $p < 0.05$; ** $p < 0.01$).

To better appreciate the potential antimicrobial effect of micelle loading, the lowest tested concentration (0.1 μM) was preferred to the highest one (1 μM). To mimic a potential disinfection protocol, aPDI was performed at increasing irradiation times from 1 to 6 h. The dark incubation for long periods of time of PS dissolved in DMSO did not cause any alteration in bacterial viability, as shown in Figure 5. On the other hand, the irradiation of PS dissolved in DMSO showed a dose-dependent light killing rate; upon 3 h and longer time of irradiation (≥26.1 J/cm^2), the highest killing rate (limit of detection of the system) was reached. The antimicrobial effect of micelles was ruled out both under dark incubation and upon irradiation.

Interestingly, the encapsulation of PS in micelles did not impair the desired photoinactivation; a decrease of *S. aureus* viability was observed, even if to a lesser extent than what was elicited by PS dissolved in DMSO. The kinetics of photoinactivation induced by micelle-loaded PSs was like that observed with BODIPY in DMSO, reaching the highest rate (−5 log unit) after a 3 h and longer period of irradiation. A bacterial sample treated with empty micelles was also incubated in the dark or irradiated for 6 h, and no cellular concentration change was observed with respect to untreated cells (data not shown). Indeed, it is important to rule out any bacterial agglutinating effect of micelles that could overestimate the antimicrobial effect.

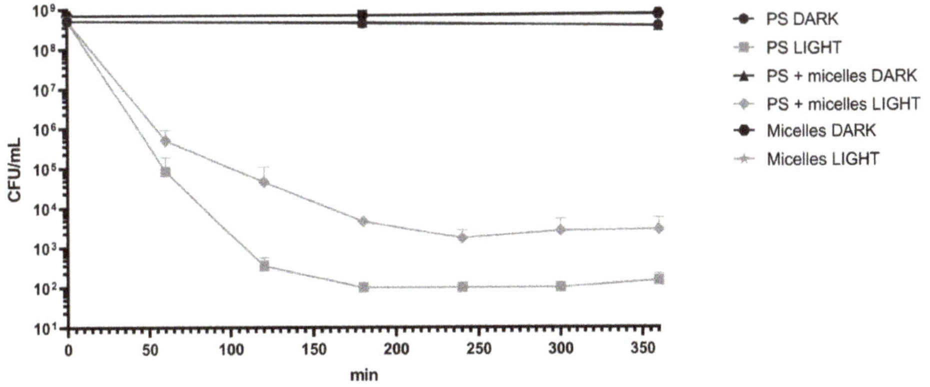

Figure 5. Photodynamic inactivation of *Staphylococcus aureus* with 0.1 μM BODIPY dissolved in DMSO or encapsulated in micelles. The irradiation with light at 520 nm was delivered up to 6 h, and the effect of micelle-encapsulated BODIPY and of BODIPY was checked every 60 min. Cell viability of control samples was checked upon 3 and 6 h of dark incubation or irradiation. Data are expressed as CFU/mL ± standard deviations.

The latter experiments showed that the delivery system of micelles worked very well; a decrease of 5 log unit from bacterial samples at a very high concentration (~10^9 CFU/mL) is an important microbiological goal. Furthermore, this delivery system represents a relevant alternative to DMSO as a carrier for BODIPY surface distribution, as DMSO has several drawbacks, such as environmental toxicity and difficult removal from surfaces.

To evaluate if the antimicrobial effect of the system could be observed upon coating, the following experiment was planned, as described in detail in Material and Methods. A coverslip glass was chosen as a surface to coat by spraying on it a water solution of micelles loaded with BODIPY and, thereafter, to inoculate with *S. aureus*. Upon overnight incubation, the irradiation (17.4 J/cm^2) by light at 520 nm caused clear alterations of cell integrity that were highlighted by confocal analysis. As can be appreciated in Figure 6A, the bacteria of the control sample showed the typical morphology of well-defined cocci and with a diameter of ~1–1.5 μm. In the enlarged image, it is possible to recognize couples of cells very close one to another, as expected, at the final step of the binary fission process. This is in accordance with viable and physiological undamaged bacteria. On the other hand, the distribution of the chosen fluorophore inside microorganisms inoculated on micelle coated surface was very different, as shown in Figure 6B. In this case, a diffuse signal of a fluorophore was detectable. It was not possible to identify the bacterial morphology, and, in addition, the dimensions of the photo-inactivated cells were smaller than those of the untreated ones. Furthermore, there was the appearance of clusters of cells not visible in the control sample. Induction of cell aggregation has already been observed in *S. aureus* strains [48] under adverse growth conditions, such as the presence of antibiotics in sub-toxic concentrations. The cluster formation thus represents a defensive strategy of bacteria to resist and is associated with the production of polysaccharides that favor cellular association and confer physical barrier properties to chemical agents. In the aggregate, there are dead and living cells, so that the polysaccharides and the presence of dead cells could explain the diffuse fluorescence we observed only in the case of coverslip glass coated with loaded micelles (Figure 6B).

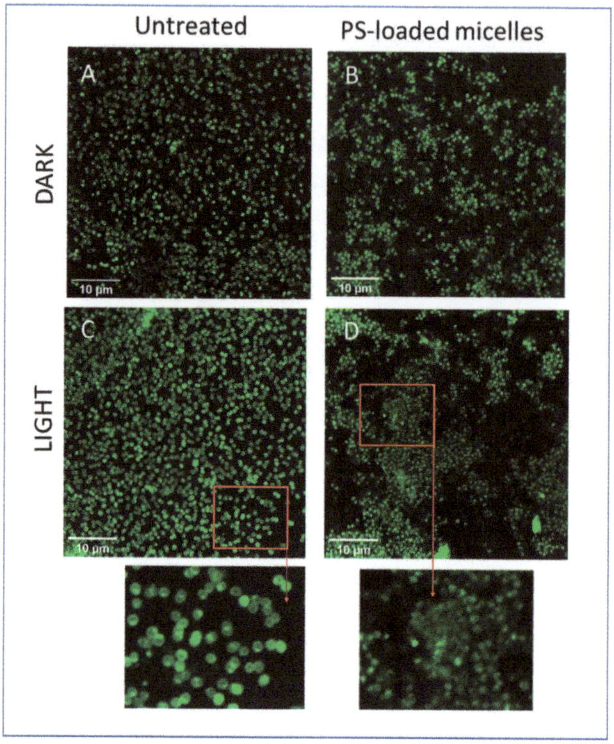

Figure 6. Photodynamic inactivation of *Staphylococcus aureus* grown on coverslip glasses. *S. aureus* cells were inoculated on coverslip glass (**A**,**C**) or on coverslip glass where a volume of 200 µL of PS-loaded micelle (0.8 µM) solutions was sprayed on (**B**,**D**). Upon overnight growth, cells were irradiated under light at 520 nm for 2 h (**C**,**D**) or dark incubated (**A**,**B**). The fluorochrome was added 30 m before confocal analysis. The images were acquired through a 63X objective lens and a 488 nm laser.

Taken together, these observations suggest that the coated surface changed greatly the appearance and most likely compromised the physiological status of the microbial population.

The combination of micelles and PSs seems very promising in the antimicrobial field and suggests further improvements. Since polymers with intrinsic photosensitizing activity have been investigated [49,50], a novel potential antimicrobial technique could rely on synergistic activity of photoactivable micellar components combined with PSs. This approach could take advantage of the antimicrobial activity of both involved parts, the container and the content.

4. Conclusions

In this study, we report a strategy to decontaminate surfaces from *S. aureus*, one of the more difficult human pathogens to eradicate, by a coating made of micelles loaded with the neutral BODIPY 2,6-diiodo-1,3,5,7-tetramethyl-8-(2,6-dichlorophenyl)-4,4′-difluoroboradiazaindacene. Micelles formed from mPEG–(PLLA)$_n$ with mPEG 5kDa and n = 2 or 4 were found to be the most thermodynamically stable among those obtained by self-assembling several copolymers based on mPEG and PLA branches, the latter both atactic and stereoregular. The higher stability of micelles obtained from the stereoregular branched copolymers might be ascribed to the stronger hydrophobic interaction among the PLLA arms with respect to the corresponding atactic copolymer, and for the presence of more than one PLLA chain per mPEG block. Release of BODIPY from micelles obtained by self-assembling the stereoregular branched copolymers in solution showed a very different behavior depending on the copolymer structure. Interestingly, the most stable supramolecular system, based on mPEG–$(PLLA)_2$, was able to release faster, probably for a different localization of the dye within the micelles. PSs are likely placed in the outer part of the hydrophobic core and in proximity to the hydrophilic corona. The faster releasing system was the one tested in the antimicrobial activity, and two significant results emerged from experimental data. For the first time it was demonstrated that a neutral BODIPY, after irradiation, is active against *S. aureus* even at micromolar concentrations, and that the system releasing such a BODIPY induces an important decrease of 5 log unit in the presence of a high concentration (~10^9 CFU/mL) of bacteria.

Overall, the supramolecular BODIPY releasing system proposed in this paper as a coating for surface decontamination allows for the overcoming of several complications deriving from the use of photodynamic active coating containing PSs, meaning ROS diffusion through the polymer matrix responsible for their partial deactivation or oxygen matrix permeability required to produce $^1O_2$.

More in general, such a system is highly versatile, since in principle it is possible to modulate PS release by properly choosing the structure of copolymer forming micelles and the nature of PSs.

Author Contributions: E.C. and V.T.O. equally contributed. M.C.M. and E.M. are students in "Life Science and Biotechnology"; C.S. is a student in "Chemistry and Environmental Chemistry". Conceptualization, E.C., V.T.O., L.I., D.P. and G.V.; Investigation, M.C.M. and E.M.; Methodology, E.C., V.T.O. and L.I.; Supervision, E.C., V.T.O. and L.I.; Validation, M.C.M. and E.M.; Visualization, C.S., D.P. and G.V.; Writing—original draft, L.I.; Writing—review and editing, L.I. All authors have read and agreed to the published version of the manuscript.

Funding: This research received no external funding.

Institutional Review Board Statement: Not applicable.

Informed Consent Statement: Not applicable.

Data Availability Statement: Data is contained within the article.

Acknowledgments: D.P. acknowledges funding from the "Fondo di Ateneo per la Ricerca di Base (FARB 2017)". L.I. acknowledges funding from the "Fondo di Ateneo per la Ricerca (FAR 2018)".

Conflicts of Interest: The authors declare no conflict of interest.

References

1. Russel, A.D. Bacterial resistance to disinfectant. *J. Infect. Prev.* **2002**, *3*, 22–24. [CrossRef]
2. Campoccia, D.; Montanaro, L.; Arciola, C.R. A review of the biomaterial technologies for infection-resistant surfaces. *Biomaterials* **2013**, *34*, 8533–8554. [CrossRef] [PubMed]
3. Tamanna, T.; Bulitta, J.B.; Landersdorfer, C.B.; Cashin, V.; Yu, A. Stability and controlled antibiotic release from thin films embedded with antibiotic loaded mesoporous silica nanoparticles. *RSC Adv.* **2015**, *5*, 107839–107846. [CrossRef]
4. Banfi, S.; Caruso, E.; Orlandi, V.T.; Barbieri, P.; Cavallari, S.; Viganò, P.; Clerici, P.; Chiodaroli, L. Antibacterial activity of leaf extracts from Combretum micranthum and Guiera senegalensis (Combretaceae). *Res. J. Microbiol.* **2014**, *9*, 66–81. [CrossRef]
5. Takahashi, H.; Caputo, G.A.; Vemparala, S.; Kuroda, K. Synthetic Random Copolymers as a Molecular Platform to Mimic Host-Defense Antimicrobial Peptides. *Bioconjug. Chem.* **2017**, *28*, 1340–1350. [CrossRef]
6. Caruso, E.; Malacarne, M.C.; Banfi, S.; Gariboldi, M.B.; Orlandi, V.T. Cationic diarylporphyrins: In vitro versatile anticancer and antibacterial photosensitizers. *J. Photochem. Photobiol. B* **2019**, *197*, 111548. [CrossRef]
7. Cieplik, F.; Deng, D.; Crielaard, W.; Buchalla, W.; Hellwig, E.; Al-Ahmad, A.; Maisch, T. Antimicrobial photodynamic therapy—what we know and what we don't. *Crit. Rev. Microbiol.* **2018**, *44*, 571–589. [CrossRef] [PubMed]
8. Hamblin, M.R.; Abrahamse, H. Can light-based approaches overcome antimicrobial resistance? *Drug Dev. Res.* **2018**, *80*, 48–67. [CrossRef]
9. Gerola, A.P.; Costa, P.F.A.; de Morais, F.A.P.; Tsubone, T.M.; Caleare, A.O.; Nakamura, C.V.; Brunaldi, K.; Caetano, W.; Kimura, E.; Hioka, N. Liposome and polymeric micelle-based delivery systems for chlorophylls: Photodamage effects on Staphylococcus aureus. *Colloids Surfaces B Biointerfaces* **2019**, *177*, 487–495. [CrossRef]
10. Gao, Y.; Wang, J.; Hu, D.; Deng, Y.; Chen, T.; Jin, Q.; Ji, J. Bacteria-Targeted Supramolecular Photosensitizer Delivery Vehicles for Photodynamic Ablation Against Biofilms. *Macromol. Rapid Commun.* **2018**, *40*, 1800763. [CrossRef] [PubMed]

11. McCoy, P.C.; O'Neil, E.J.; Cowley, J.F.; Carson, L.; De Baroid, A.T.; Gdowski, G.T.; Gorma, S.P.; Jones, D.S. Photodynamic Antimicrobial Polymers for Infection. *PLoS ONE* **2014**, *9*, e108500. [CrossRef]
12. Qin, Y.; Chen, L.L.; Pu, W.; Liu, P.; Liu, S.X.; Li, Y.; Liu, X.L.; Lu, Z.X.; Zheng, L.Y.; Cao, Q.E. A hydrogel directly assembled from a copper metal-organic polyhedron for antimicrobial application. *Chem. Commun.* **2019**, *55*, 2206–2209. [CrossRef]
13. Henke, P.; Kirakci, K.; Kubat, P.; Fraiberk, M.; Forstovà, J.; Mosinger, J. Antibacterial, antiviral, and oxygen-sensing nanoparticles prepared from electrospun materials. *ACS Appl. Mater. Interfaces* **2016**, *8*, 25127–25136. [CrossRef] [PubMed]
14. Zhang, F.; Zhang, C.L.; Peng, H.Y.; Cong, H.P.; Qian, H.S. Near-infrared photocatalytic upconversion nanoparticles/TiO_2 nanofibers assembled in large scale by electrospinning. *Part. Part. Syst. Charact.* **2016**, *33*, 248–253. [CrossRef]
15. Dolansky, J.; Henke, P.; Malà, Z.; Zàrskà, L.; Kubàt, P.; Mosinger, J. Antibacterial nitric-oxide- and singlet oxygen-releasing polystyrene nanoparticles responsive to light and temperature triggers. *Nanoscale* **2018**, *10*, 2639–2648. [CrossRef] [PubMed]
16. Thomas, D.D.; Liu, X.Q.; Kantrow, S.P.; Lancaster, J.R. The biological lifetime of nitric oxide: Implications for the perivascular dynamics of NO and O_2. *Proc. Natl. Acad. Sci. USA* **2001**, *98*, 355–360. [CrossRef] [PubMed]
17. Bregnhoj, M.; Westberg, M.; Jensen, F.; Ogilby, P.R. Solvent-dependent singlet oxygen lifetimes: Temperature effects implicate tunneling and charge-transfer interactions. *Phys. Chem. Chem. Phys.* **2016**, *18*, 22946–22961. [CrossRef]
18. Felgenträger, A.; Maisch, T.; Späth, A.; Schröder, J.A.; Bäumler, W. Singlet oxygen generation in porphyrin-doped polymeric surface coating enables antimicrobial effects on *Staphylococcus aureus*. *Phys. Chem. Chem. Phys.* **2014**, *16*, 20598–20607. [CrossRef]
19. Izzo, L.; Pappalardo, D. "Tree-Shaped" Copolymers Based on Poly(ethylene glycol) and Atactic or Isotactic Polylactides: Synthesis and Characterization. *Macromol. Chem. Phys.* **2010**, *211*, 2171–2178. [CrossRef]
20. Garofalo, C.; Capuano, G.; Sottile, R.; Tallerico, R.; Adami, R.; Reverchon, E.; Carbone, E.; Izzo, L.; Pappalardo, D. Different Insight into Amphiphilic PEG-PLA Copolymers: Influence of Macromolecular Architecture on the Micelle Formation and Cellular Uptake. *Biomacromolecules* **2014**, *15*, 403–415. [CrossRef]
21. Adami, R.; Liparoti, S.; Izzo, L.; Pappalardo, D.; Reverchon, E. PLA-PEG copolymers micronization by Supercritical Assisted Atomization. *J. Supercrit. Fluids* **2012**, *72*, 15–21. [CrossRef]
22. Caruso, E.; Banfi, S.; Barbieri, P.; Leva, B.; Orlandi, V.T. Synthesis and antibacterial activity of novel cationic BODIPY photosensitizers. *J. Photochem. Photobiol. B* **2012**, *114*, 44–51. [CrossRef] [PubMed]
23. Orlandi, V.T.; Rybtke, M.; Caruso, E.; Banfi, S.; Tolker-Nielsen, T.; Barbieri, P. Antimicrobial and anti-biofilm effect of a novel BODIPY photosensitizer against Pseudomonas aeruginosa PAO1. *Biofuling* **2014**, *30*, 883–891. [CrossRef] [PubMed]
24. Caruso, E.; Ferrara, S.; Ferruti, P.; Manfredi, A.; Ranucci, E.; Orlandi, V.T. Enhanced photoinduced antibacterial activity of a BODIPY photosensitizer in the presence of polyamidoamines. *Lasers Med. Sci.* **2018**, *33*, 1401–1407. [CrossRef] [PubMed]
25. Banfi, S.; Nasini, G.; Zaza, S.; Caruso, E. Synthesis and Photo-Physical properties of a series of BODIPY dyes. *Tetrahedron* **2013**, *69*, 4845–4856. [CrossRef]
26. Orlandi, V.T.; Martegani, E.; Bolognese, F.; Trivellin, N.; Paldrychova, M.; Matatkova, O.; Baj, A.; Caruso, E. Photodynamic therapy by diaryl-porphyrins to control Candida albicans growth. *Cosmetics* **2020**, *7*, 31. [CrossRef]
27. Meiners, J.C.; Quintel-Ritzi, A.; Mlynek, J.; Elbs, H.; Kraisch, G. Adsorption of block-copolymer micelles from a selective solvent. *Macromolecules* **1997**, *30*, 4945–4951. [CrossRef]
28. Farinha, J.P.S.; D'Oliveira, J.M.R.; Martinoho, J.M.; Xu, R.; Winnik, M.A. Structure in tethered chains: Polymeric micelles and chains anchored on polystyrene latex spheres. *Langmuir* **1998**, *14*, 2291–2296. [CrossRef]
29. Li, Y.; Pi, Q.; You, H.; Li, J.; Wang, P.; Yang, X.; Wu, Y. A smart multi-functional coating based on anti-pathogen micelles tethered with copper nanoparticles via a biosynthesis method using L-vitamin C. *RSC Adv.* **2018**, *8*, 18272–18283. [CrossRef]
30. Webber, S.E. Polymer micelles: An example of self-assembling polymers. *J. Phys. Chem. B* **1998**, *102*, 2618–2626. [CrossRef]
31. Karymov, M.A.; Prochazka, K.; Mendenhall, J.M.; Martin, T.J.; Munk, P.; Webber, S.E. Chemical Attachment of Polystyrene-*block*-poly(methacrylic acid) Micelles on a Silicon Nitride Surface. *Langmuir* **1996**, *12*, 4748–4753. [CrossRef]
32. Emoto, K.; Nagasaki, Y.; Kataoka, K. Coating of Surfaces with Stabilized Reactive Micelles from Poly(ethylene glycol)-Poly(DL-lactic acid) Block Copolymer. *Langmuir* **1999**, *15*, 5212–5218. [CrossRef]
33. Izzo, L.; Gorrasi, G.; Sorrentino, A.; Tagliabue, A.; Mella, M. Controlling Drug Release of Anti-inflammatory Molecules Through a pH-Sensitive, Bactericidal Polymer Matrix: Towards a Synergic and Combined Therapy. *Lect. Notes Bioeng.* **2019**, 151–163. [CrossRef]
34. Mella, M.; Tagliabue, A.; Mollica, L.; Izzo, L. Monte Carlo study on the effect on macroion charge distribution on the ionization and adsorption of weak polyelectrolytes and concurrent counterion release. *J. Colloids Interface Sci.* **2020**, *560*, 667–680. [CrossRef]
35. Tagliabue, A.; Izzo, L.; Mella, M. Impact of charge correlation, chain rigidity, and chemical specific interactions on the behaviour of weak polyelectrolytes in solution. *J. Phys. Chem. B* **2019**, *123*, 8872–8888. [CrossRef] [PubMed]
36. Izzo, L.; Matrella, S.; Mella, M.; Benvenuto, G.; Vigliotta, G. *Escherichia coli* as a Model for the Description of the Antimicrobial Mechanism of a Cationic Polymer Surface: Cellular Target and Bacterial Contrast Response. *ACS Appl. Mat. Interfaces* **2019**, *11*, 15332–15343. [CrossRef]
37. Barrella, M.C.; Di Capua, A.; Adami, R.; Reverchon, E.; Mella, M.; Izzo, L. Impact of intermolecular drug-copolymer interactions on size and drug release kinetics from pH-responsive polymersomes. *Supramol. Chem.* **2017**, *29*, 796–807. [CrossRef]
38. Villani, S.; Adami, A.; Reverchon, E.; Ferretti, A.M.; Ponti, A.; Lepretti, M.; Caputo, I.; Izzo, L. pH-sensitive polymersomes: Controlling swelling via copolymer structure and chemical composition. *J. Drug Target.* **2017**, *25*, 899–909. [CrossRef]

39. Tagliabue, A.; Izzo, L.; Mella, M. Out of equilibrium self-assembly of Janus nanoparticles: Steering it from disordered amorphous to 2D patterned aggregates. *Langmuir* **2016**, *32*, 12934–12946. [CrossRef] [PubMed]
40. Matrella, S.; Vitiello, C.; Mella, M.; Vigliotta, G.; Izzo, L. The Role of Charge Density and Hydrophobicity on the Biocidal Properties of Self-Protonable Polymeric Materials. *Macromol. Biosci.* **2015**, *15*, 927–940. [CrossRef]
41. Mella, M.; Mollica, L.; Izzo, L. Influence of charged intramolecular hydrogen bonds in weak polyelectrolytes: A Monte Carlo study of flexible and extendible polymeric chains in solution and near charged spheres. *J. Polymer Sci. Part B Polymer Phys.* **2015**, *53*, 650–663. [CrossRef]
42. Izzo, L.; Gorrasi, G. Effect of molecular architecture on physical properties of tree-shaped and star-shaped poly(methylmethacrylate)-based copolymers. *J. Macromol. Sci. B* **2014**, *53*, 474–485. [CrossRef]
43. Vigliotta, G.; Mella, M.; Rega, D.; Izzo, L. Modulating antimicrobial activity by synthesis: Dendritic copolymers based on nonquaternized 2-(dimethylamino)ethyl methacrylate by Cu-mediated ATRP. *Biomacromolecules* **2012**, *13*, 833–841. [CrossRef]
44. Glavas, L.; Olsén, P.; Odelius, K.; Albertsson, A.-C. Achieving Micelle Control through Core Crystallinity. *Biomacromolecules* **2013**, *14*, 4150–4156. [CrossRef] [PubMed]
45. Caruso, E.; Gariboldi, M.B.; Sangion, A.; Gramatica, P.; Banfi, S. Synthesis, photodynamic activity, and quantitative structure-activity relationship modelling of a series of BODIPYs. *J. Photochem. Photobiol. B* **2017**, *167*, 269–281. [CrossRef]
46. Álvarez, A.; Fernández, L.; Gutiérrez, D.; Iglesias, B.; Rodríguez, A.; García, P. Methicillin-Resistant Staphylococcus aureus in Hospitals: Latest Trends and Treatments Based on Bacteriophages. *J. Clin. Microbiol.* **2019**, *57*, e01006-19. [CrossRef]
47. Ghorbani, J.; Rahban, B.; Aghamiri, S.; Teymouri, A.; Bahador, A. Photosensitizers in antibacterial photodynamic therapy: An overview. *Laser Ther.* **2018**, *27*, 293–302. [CrossRef]
48. Haaber, J.; Thorup Cohn, M.; Frees, D.; Andersen, T.J.; Ingmer, H. Planktonic Aggregates of *Staphylococcus aureus* Protect against Common Antibiotics. *PLoS ONE* **2012**, *7*, e41075. [CrossRef]
49. Zhao, Q.; Li, J.; Zhang, X.; Li, Z.; Tang, Y. Cationic Oligo(thiophene ethynylene) with Broad-Spectrum and High Antibacterial Efficiency under White Light and Specific Biocidal Activity against S. aureus in Dark. *ACS Appl. Mater. Interfaces* **2016**, *8*, 1019–1024. [CrossRef] [PubMed]
50. Zhou, Z.; Ergene, C.; Lee, J.Y.; Shirley, D.J.; Carone, B.R.; Caputo, G.A.; Palermo, E.F. Sequence and Dispersity Are Determinants of Photodynamic Antibacterial Activity Exerted by Peptidomimetic Oligo(thiophene)s. *ACS Appl. Mater. Interfaces* **2019**, *11*, 1896–1906. [CrossRef] [PubMed]

coatings

MDPI

Article

Antimicrobial Peptides Grafted onto a Plasma Polymer Interlayer Platform: Performance upon Extended Bacteria Challenge

Stefani S. Griesser, Marek Jasieniak, Krasimir Vasilev and Hans J. Griesser *

Future Industries Institute, University of South Australia, Mawson Lakes, SA 5095, Australia; stefanigriesser@gmail.com (S.S.G.); marek.jasieniak@unisa.edu.au (M.J.); krasimir.vasilev@unisa.edu.au (K.V.)
* Correspondence: hans.griesser@unisa.edu.au

Abstract: To combat infections on biomedical devices, antimicrobial coatings have attracted considerable attention, including coatings comprising naturally occurring antimicrobial peptides (AMPs). In this study the aim was to explore performance upon extended challenge by bacteria growing in media above samples. The AMPs LL37, Magainin 2, and Parasin 1 were selected on the basis of well-known membrane disruption activity in solution and were covalently grafted onto a plasma polymer platform, which enables application of this multilayer coating strategy to a wide range of biomaterials. Detailed surface analyses were performed to verify the intended outcomes of the coating sequence. Samples were challenged by incubation in bacterial growth media for 5 and 20 h. Compared with the control plasma polymer surface, all three grafted AMP coatings showed considerable reductions in bacterial colonization even at the high bacterial challenge of initial seeding at 1×10^7 CFU, but there were increasing numbers of dead bacteria attached to the surface. All three grafted AMP coatings were found to be non-toxic to primary fibroblasts. These coatings thus could be useful to produce antibacterial surface coatings for biomaterials, though possible consequences arising from the presence of dead bacteria need to be studied further, and compared to non-fouling coatings that avoid attachment of dead bacteria.

Keywords: antibacterial coating; antimicrobial peptide; plasma polymer; LL 37; Magainin; Parasin; bacterial attachment

Citation: Griesser, S.S.; Jasieniak, M.; Vasilev, K.; Griesser, H.J. Antimicrobial Peptides Grafted onto a Plasma Polymer Interlayer Platform: Performance upon Extended Bacterial Challenge. *Coatings* **2021**, *11*, 68. https://doi.org/10.3390/coatings11010068

Received: 4 December 2020
Accepted: 28 December 2020
Published: 8 January 2021

1. Introduction

The occurrence of infections on biomedical devices such as catheters, hip and knee implants, contact lenses, and many others, is a major challenge in healthcare leading to patient morbidity and mortality, and enormous added costs [1–5]. Many of these infections arise from the ability of bacterial and fungal pathogens to attach to surfaces of devices, proliferate, and form biofilms [4,5]. Once formed, such infectious biofilms are difficult to eradicate [4,5]. Accordingly, there has been much interest in research aimed at developing surfaces and coatings that can prevent device-associated infections, by stopping either microbial attachment or the ability of attached microbes to convert to the biofilm-forming phenotype [6–9].

One class of molecules that has attracted considerable interest for the development of antimicrobial coatings is that of antimicrobial peptides (AMPs) [10–15]. AMPs are part of the innate immune system and are involved in the first line of defense against bacterial invasion for all multicellular organisms [10,11]. AMPs have been isolated from a wide variety of animals, plants, bacteria, fungi and virus [11]. They function as both antimicrobial agents and modulators of the immune system [14,16]. Although they are highly diverse, they have three characteristics that are shared by almost all known AMPs: a relatively small size (10–40 amino acids), a highly cationic character, and an amphipathic nature [17]. AMPs offer many significant potential advantages in that they have broad-spectrum activity across a broad range of Gram-positive and Gram-negative bacteria, including drug-resistant strains, and are also active against fungi [11,18]. Many AMPs target bacterial membrane function and stability, rather than specific protein binding sites [18]. This makes them highly advantageous because they can kill microbes in growing and non-growing states and in dormancy, and do not induce resistance [13].

A natural extension to the extensive research in AMPs is their application to solid surfaces of materials and medical devices via various chemistries to deter bacterial surface colonisation and biofilm formation [17,19–21]. To retain AMPs on surfaces of biomedical devices in biological environments, they must be bound covalently ("grafted") to the biomaterial surface. There are many reports on grafting of AMPs onto various materials by various interfacial linking chemistries, as discussed in recent reviews [22–24]. In this study, we have utilized the approach of using a plasma polymer interlayer bearing surface aldehyde groups that can react with amine groups of AMPs to form an interfacial covalent bond [25]. The attraction of using a plasma polymer interlayer is that identical plasma polymers can facilely be deposited onto a wide range of materials and devices, and hence our coating strategy is generically applicable to a wide range of potential products [7,25–27]. Other surface modification techniques are typically limited to specific substrate materials. For example, Layer by Layer deposition requires a charged surface, whilst Self Assembled Monolayers require metallic surfaces in the case of thiols or a silica surface in the case of silanes [28]. Notably, plasma polymers are deposited from the vapour phase of a carefully chosen precursor and thus do not involve the use of solvents. As a consequence, there are no requirements for waste solvent treatment or pollution to the environment. Lastly, the (electrical) energy required for plasma polymerisation can be potentially generated from purely sustainable sources such as solar or wind.

Previous studies with AMPs grafted onto surfaces using other chemistries have shown high effectiveness [17,19,20]. However, typically relatively short inoculation times and moderate bacterial challenges (numbers in solution) were employed. Accordingly, in this study the main focus was on studying effects arising after extended periods of challenging samples with high loads of bacteria, after showing that an aldehyde plasma polymer is suitable for grafting AMPs in an active conformation. Three well-characterized AMPs known to have potent antibacterial properties in solution were used. We found that whilst effective initially at resisting bacterial colonization, eventually the coatings became colonized by increasing numbers of dead and live bacteria. On one coating, once dead bacteria had accumulated this then allowed live bacteria to attach on top of the dead bacterial layer.

2. Materials and Methods

2.1. AMPs

For this work, three well-known, representative AMPs from different classes were selected, on the basis of suitable properties detailed in the Discussion section: LL37(134–170), Magainin 2, and Parasin 1. The amino acid compositions are: LL37 (MW 4493.37 Da): H-Leu-Leu-Gly-Asp-Phe-Phe-Arg-Lys-Ser-Lys-Glu-Lys-Ile-Gly-Lys-Glu-Phe-Lys-Arg-Ile-Val-Gln-Arg-Ile-Lys-Asp-Phe-Leu-Arg-Asn-Leu-Val-Pro-Arg-Thr-Glu-Ser-OH; Magainin 2 (MW 2466.95 Da): H-Gly-Ile-gly-Lys-Phe-Leu-His-Ser-Ala-Lys-Lys-Phe-Gly-Lys-Ala-Phe-Val-Gly-Glu-Ile-Met-Asn-Ser-OH; and Parasin 1 (MW 2000.3 Da): H-Lys-Gly-Arg-Gly-Lys-Gln-Gly-Gly-Lys-Val-Arg-Ala-Lys-Ala-Lys-Thr-Arg-Ser-Ser-OH. They were purchased from GL Biochem, Shanghai, China, with > 97% purity.

2.2. Grafting Methodology

The reaction scheme for covalent grafting onto the aldehyde plasma polymer (ALDpp) surface via reductive amination is shown schematically in Scheme 1.

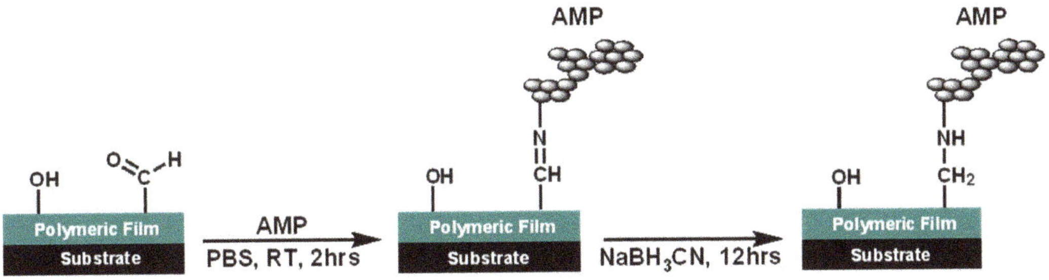

Scheme 1. Immobilisation of AMPs via reductive amination onto ALDpp.

The ALDpp interlayer was deposited in a plasma system described previously [29] and using plasma conditions optimized in a previous study [30] to ~21 nm thickness, as measured by ellipsometry on Si wafer substrates, onto several substrates to suit intended tests and demonstrate application to different material surfaces: silicon wafers (MMRC, Malvern, Australia), Thermanox coverslips (GL083, ProSciTech, Brisbane, Australia), ibidi well plates (81201 and 81821, DKSH, Hallam, Australia), and polystyrene slides (ProSciTech, Brisbane, Australia). FlexiPERM 12 reusable masks (Sarstedt, Mawson Lakes, Australia) were used to define areas for coatings. ALDpp coated samples were placed in sterile solutions of 0.1 mg/mL AMPs in PBS for 2 h at 21 °C, followed by addition of an equal volume of sodium cyanoborohydride, 1 mg/mL, and letting reduction proceed at 4 °C for 12 h. Samples were then soaked and rinsed six times over 2 h with sterile PBS. The final rinse was collected and tested with bacteria to check that no peptide detached into solution and therefore the results were indeed from covalently attached peptides only.

2.3. Surface Analysis

Samples were analyzed using X-ray photoelectron spectroscopy (XPS) and Time-of-Flight Secondary Ion Mass Spectrometry (ToF-SIMS). XPS was performed with a Kratos AXIS Ultra DLD spectrometer, using monochromatic AlK_{α} radiation ($h\nu$ = 1486.7 eV) and a magnetically confined charge compensation system. Spectra were recorded using an acceleration voltage of 15 keV at a power of 225 W. Survey spectra were collected with a pass energy of 160 eV and an analysis area of 300 × 700 μm^2. High-resolution spectra were obtained using a 20 eV pass energy and an analysis area of ~300 × 700 μm^2. Data analysis was performed with CasaXPS software (Casa Software Ltd.). All binding energies were referenced to the neutral component of the C 1 s peak at 285.0 eV. Core level envelopes were curve fitted with the minimum number of mixed Gaussian–Lorentzian component profiles. The Gaussian–Lorentzian mixing ratio (typically 30% Lorentzian and 70% Gaussian functions); the full width at half maximum, and the positions and intensities of peaks were left unconstrained to result in a best fit.

ToF-SIMS measurements were performed with a PHI TRIFT V nanoTOF instrument (PHI Electronics Ltd., Chanhassen, MN, USA), with a 30 keV, pulsed primary $^{197}Au^+$ ion beam and dual beam charge neutralisation using a combination of low-energy argon ions (up to 10 eV) and electrons (up to 25 eV). Positive mass axis calibration was done with CH_3^+, $C_2H_5^+$ and $C_3H_7^+$. Spectra were acquired in the bunched mode for 60 s from an area of 100 μm × 100 μm. The corresponding total primary ion dose was less than 1×10^{12} ions cm^{-2}, and thus met the conditions of the static SIMS regime [31]. A mass resolution m/Δm of > 7000 at nominal m/z = 27 amu ($C_2H_3^+$) was typically achieved. Some samples were characterised by multiple positive ion mass spectra, collected from sample areas that did not overlap. All peaks not obscured by overlaps in the amu range 2 to 175 were used in Principal Component Analysis (PCA) calculations. Peak intensities were normalized to the total intensity of all peaks. Multiple mass spectra were processed by PCA [32]. using PLS_Toolbox version 3.0 (Eigenvector Research, Inc., Manson, WA, USA) along with MATLAB software v. 6.5 (MathWorks Inc., Natick, MA, USA).

2.4. *Bacterial Testing*

Bacteria (*S. epidermidis* ATCC 35984, *S. aureus* MRSA ATCC 43300, and *E. coli* ATCC 35922) were plated from frozen stock and incubated overnight at 37 °C. Two colonies were picked from plates and grown in 10 mL TSB (Oxoid, via Thermo Fisher, Thebarton, Australia) overnight at 37 °C, followed by dilution 1/100 and growth to log phase, then diluted to 1×10^6 bacteria/mL using calibrated spectroscopy measurements. Samples were immersed in 100 µL or 300 µL of bacterial solutions and left to grow for 5 or 24 h at 37 °C, followed by rinsing and analyses using visualization by the LIVE/DEAD *Bac*Light Bacterial Viability assay (Invitrogen, via Thermo Fisher, Thebarton, Australia), viable bacteria count plating, surface stamping onto agar plates, and safranin staining of biofilm and spectroscopy readings.

*Bac*Light contains two nucleic acid stains: a green fluorescent stain, SYTO 9, which is membrane permeable, and a red fluorescent stain, propidium iodide, which is membrane impermeant and should only stain cells that have compromised membranes. In principle, live bacteria are stained green and dead bacteria are stained red. It is important, however, that the *Bac*Light kit be tested in each system before use to ensure accurate scoring of live and dead cells. This was done by growing bacteria on supporting reference surfaces, and negative controls (dead bacteria) were created by treating bacteria in wells with Virkon for 2 min. The controls showed good reproducibility. In contrast, biofilm staining by safranin was not reliable because of the combination of dead and live bacteria present.

Polystyrene slides were plasma treated and then a FlexiPERM 12-well removable mask was placed on the slide; AMPs were added to the wells for grafting overnight. After rinsing, bacteria were added for incubation for various periods, followed by rinsing to remove loosely attached bacteria. The wells were treated with the LIVE/DEAD *Bac*Light stain and the mask carefully removed before microscopy examination.

2.5. *Fibroblasts Testing*

Two fibroblast cell lines were used: HFF-1 (human, ATCC SCRC-1041) and 3T3 (NIH-3T3 mouse fibroblast, ATCC CRL-1658), as well as freshly harvested primary fibroblasts from human explant skin, using a reported protocol [33]. All cell cultures were maintained according to ATCC instructions; primary cells were maintained according to [33]. Thermanox coverslip samples were placed in sterile 24-well culture plates (Nunc, Invitrogen) and prewarmed in a cell culture oven. Cells were made up to 50,000 cells per ml in medium/serum and added at 1 mL to each well; primary cells were made up to 10,000 cells per mL. Well plates were placed into a culture oven and left for 48 h. The coverslips were rinsed to remove non-adherent cells and placed on a glass slide and coverslip for immediate microscopic analysis.

3. Results

3.1. *Grafting of AMPs*

The three AMPs possess amine groups and thus should be amenable to convenient grafting from aqueous solutions onto surface aldehyde or surface epoxy groups, as reported previously for other proteins [30,34]. In this study, to effect grafting onto materials surfaces that *per se* do not contain aldehyde groups, a plasma polymer interlayer deposited from propanal (aka propionaldehyde) was utilized. Its surface has previously been shown to contain reactive aldehyde groups [30,34]. as well as hydroxyl groups that help provide a hydrophilic nature to the surface, which helps avoid denaturation of grafted proteins. However, reactions at surfaces may differ from reactions in solution and thus, as discussed by Castner and Ratner [35]. coatings need to be appropriately characterized prior to biological tests in order to ensure that biological responses can be interpreted reliably and possible incorrect inferences arising from unrecognized artefacts and contaminants are avoided. To verify that grafting had indeed occurred, XPS and ToF-SIMS analyses were performed.

XPS analysis of samples after immersion in solutions of AMPs showed substantial changes, relative to the ALDpp interlayer, in accord with expectations based on an immobilised protein layer (Table 1). The data show high surface coverage, particularly for LL37. LL37 possesses multiple amine groups, so would naturally have greater capacity for attachment than peptides with a smaller number of amine groups. Angle-dependent XPS showed an increase by ~33% in the N signal when the take-off angle was changed from 0° to 75° (relative to the surface normal). This indicates, as expected, that the peptides are on top of the ALDpp layer, as opposed to possible in-diffusion of the peptides into the ALDpp.

Table 1. Compositions determined by XPS of plasma polymers (ALDpp) on Thermanox (Th) substrates, and after grafting with antimicrobial peptides.

Sample	Concentration, at %		
	O	N	C
Th-ALDpp	10.6	0	89.4
Th-ALDpp-LL37	12.5	5.5	82.0
Th-ALDpp-Mg2	11.7	3.4	84.9
Th-ALDpp-Pa1	11.3	1.5	87.1

The component fitting (Table 2) also is in excellent agreement with expectations based on a surface-attached protein layer. In particular, the presence of a component assignable to amide C confirms that the elemental changes are due to an attached peptide layer, as opposed to possible surface contaminants such as adventitious hydrocarbons or fatty acids or amides. Representative spectra are shown in Figure 1.

Table 2. Component fitting for the XPS C 1s signals recorded with ALDpp and after grafting with antimicrobial peptides, on Thermanox substrates.

Sample	Fitted Components in C 1 s, (% of Total C)				
	C–C/H	C–N	C–O	C=O	N–C=O
Th-ALDpp	87.2	–	8.7	4.08	–
-ALDpp-LL37	67.0	11.9	11.5	–	9.6
-ALDpp-Mg2	76.8	9.5	8.0	–	6.4
-ALDpp-Pa1	81.5	7.8	7.0	–	3.8

ToF-SIMS, a technique capable of providing information on molecular structural elements, also gave spectra (Figures 2 and 3) that verified the presence of the AMPs, via characteristic peaks that could be assigned as originating from specific amino acids, as shown in Figure 2b, on the basis of published immonium ion signals [36]. No such signals appeared in spectra recorded with ALDpp samples. No contamination, particularly by silicones, was detected on any of the samples. Recording spectra on duplicate samples and on several separate areas of a sample also showed excellent reproducibility and uniformity of the coatings. Moreover, samples were washed and soaked extensively to probe whether the AMPs were indeed covalently attached; this did not lead to any measurable changes in spectra. It is noteworthy in Figure 3 that in addition to the commonly observed $C_2H_5N^+$ ion the high-resolution spectra after LL37 and Pa1 grafting also showed a signal attributable to the $CH_3N_2^+$ ion, which arises from arginine residues.

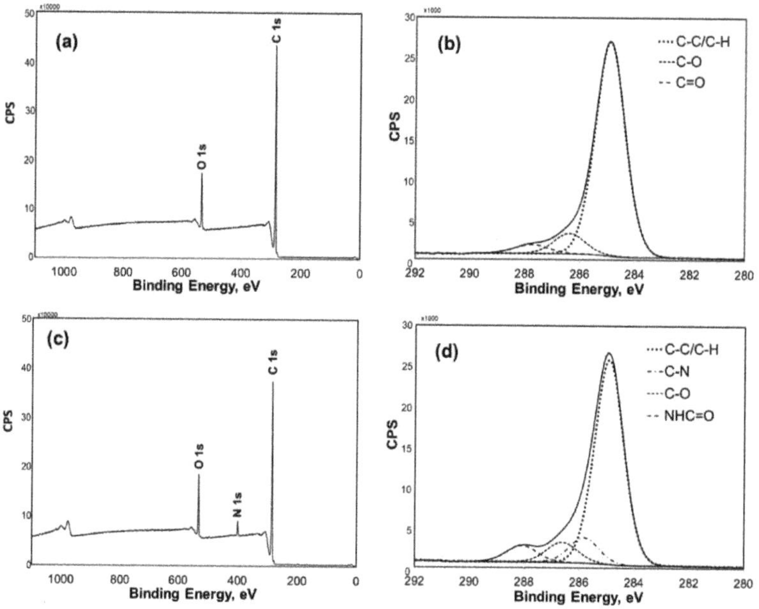

Figure 1. XPS spectra of ALDpp and ALDpp-Mg2 surfaces, on Thermanox substrates: (**a**) ALDpp, survey; (**b**) ALDpp, C 1 s; (**c**) ALDpp-Mg2, survey; (**d**) ALDpp-Mg2, C 1 s.

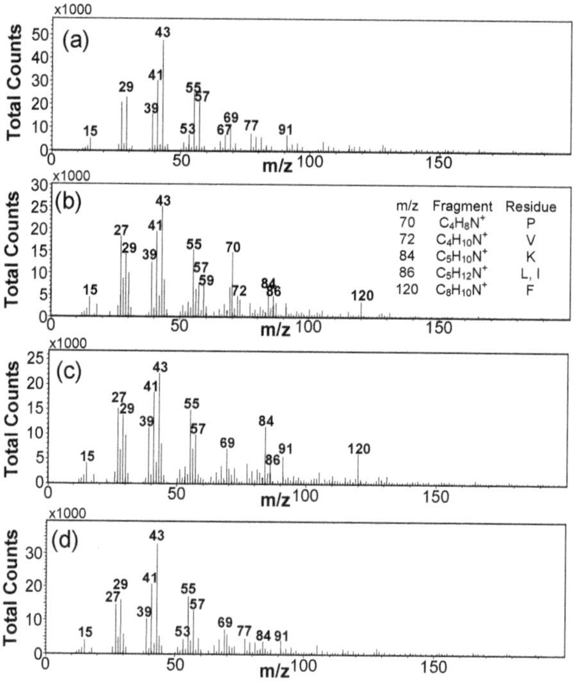

Figure 2. Positive mass spectra of ALDpp and its modifications with antimicrobial peptides: (**a**) Thermanox-ALDpp, (**b**) Th-ALDpp-LL37, (**c**) Th-ALDpp-Mg2, (**d**) Th-ALDpp-Pa1.

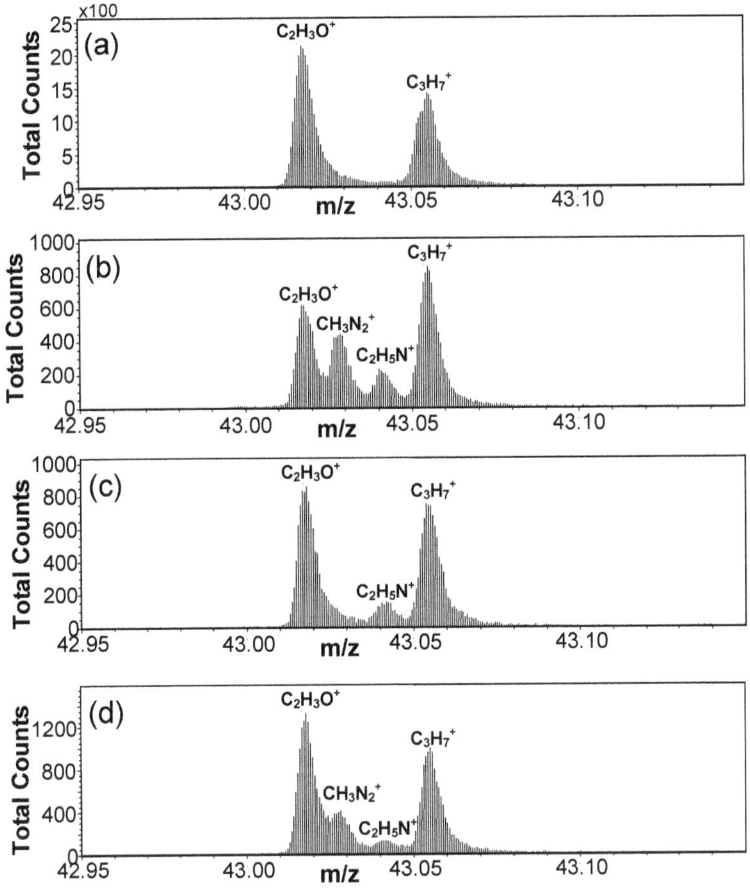

Figure 3. Positive ion mass spectra in the range of *m/z* 42.95–43.15 for ALDpp and after grafting with antimicrobial peptides: (**a**) Thermanox-ALDpp, (**b**) Th-ALDpp-LL37, (**c**) Th-ALDpp-Mg2, (**d**) th-ALDpp-Pa1.

Spectra were further processed by PCA. The scores plot on the first two principal components, which together contained 97% of the information, is shown in Figure S1 (Supplementary). The tight clustering of data recorded on separate sample areas again shows high uniformity of the surfaces. In agreement with the XPS data showing the lowest %N for parasin, the PCA analysis shows the Pa1 coating to be less distant from the ALDpp than the other two AMP graft coatings. Loadings plots, two examples of which are reproduced in Figure S2 (Supplementary), reveal the individual peaks that contribute most strongly to the differences in surface composition. Not surprisingly, the peaks that load negatively on PC1 (i.e., increase in relative intensity upon grafting of AMPs onto ALDpp) can be assigned to molecular ions that contain N. The loadings plot on PC1 in Figure S2a illustrates that along the PC1 axis the main differences arise in signals assignable to amino acids (loading negatively) and signals assignable to the underlying ALDpp, which load positively and hence are of reduced relative intensity after grafting, as expected. Interestingly, peaks assignable to the amino acid arginine are of reduced relative intensity. Its guanidine side chain should, based on chemical principles, be highly reactive with aldehyde surface groups, more so than the amino side group of lysine. The data suggest that interfacial immobilisation via reaction between arginines and surface aldehydes is an important aspect in the covalent grafting of these AMPs.

In summary, the surface analysis data clearly show that all three AMPs were successfully grafted onto the ALDpp interlayer, with LL37 grafted to the highest surface coverage and Pa1 to the lowest. Repeating these analyses after extended soaking of samples in PBS followed by rinsing gave identical results, indicating that the surface-bound AMPs were covalently grafted and thus not detachable.

3.2. Bacterial Testing

Representative optical microscopy images of bacterial growth on samples after 5 h incubation and *Bac*Light staining are shown in Figure 4. On the ALDpp, bacteria attached and grew well, while there was substantially less growth on the AMP surfaces. The images suggest that the grafted Magainin 2 coating performs best, in that it shows few bacteria either dead or alive. On the other two AMP-grafted surfaces there is clear evidence of adhering dead bacteria, as well as a significant number of live bacteria on the LL37 coating. One possible interpretation is that on grafted Magainin 2, surface-contacting bacteria are killed rapidly, before they can establish a sufficiently strong adhesive bond to the surface, whereas on the other two surfaces, some bacteria manage to attach with sufficient strength before they are killed, and thus their dead remnants then do not detach. This seems to be the case less for LL37 than for Pa 1.

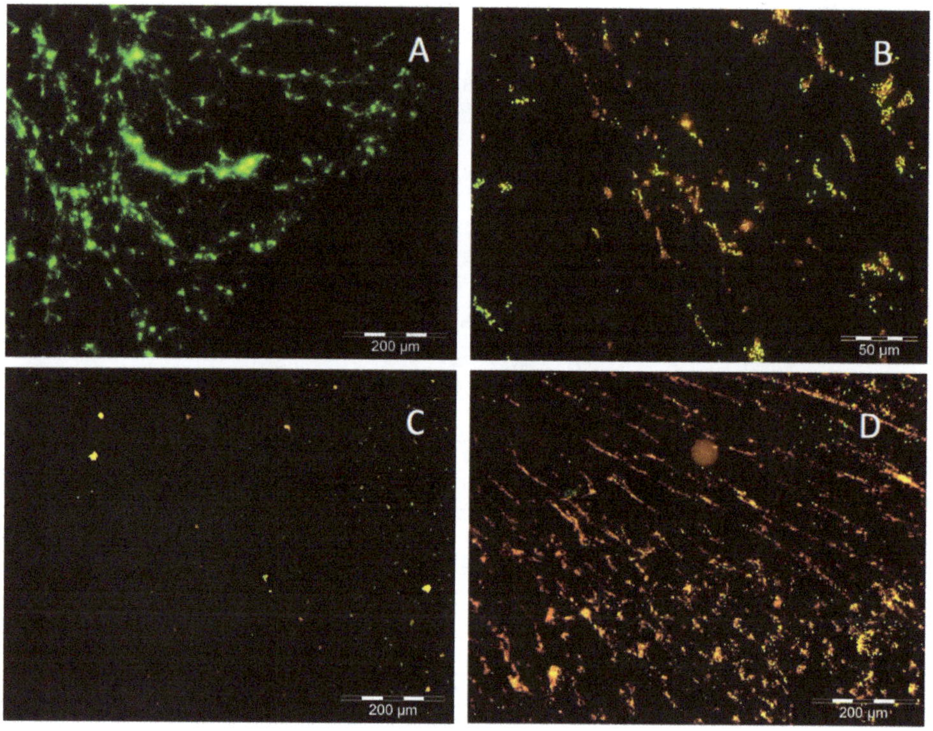

Figure 4. Stained microscopy images of *S. epidermidis* colonization after 5 h incubation, seeded at 1×10^7 CFU, on (**A**) control aldehyde plasma polymer (ALDpp), on (**B**) grafted LL37, (**C**) Magainin 2, and (**D**) Parasin 1.

For parallel samples, not stained by *Bac*Light, bacteria were removed, serially diluted, and plated on Agar. This gave colony counts of >3000 live colonies on the control ALDpp surface, ~150 colonies on LL37 grafted samples, ~70 colonies on Magainin 2 grafted samples, and ~300 colonies on Parasin 1 grafted samples. The dead and dying bacteria, which are particularly prominent on the Parasin 1 grafted coating, however, could not be quantified.

Thus, as in previous studies [17,19,20], the grafted AMPs cause a marked reduction in bacterial attachment relative to the reference polymer surface (in this case, ALDpp, which is itself not particularly adhesive for bacteria). However, whilst earlier work often has recorded only the short-term benefits, the observation of attached dead bacteria raises the question as to what the longer-duration consequences might be.

Accordingly, bacterial attachment and growth was studied over longer time frames, keeping samples in the original bacterial growth media solutions, which means that bacterial numbers were increasing steadily and thereby continuously upping the challenges on the coatings' ability to resist bacterial colonization. Representative images recorded with stained samples after 20 h of exposure to *S. epidermidis* solution are shown in Figure 5.

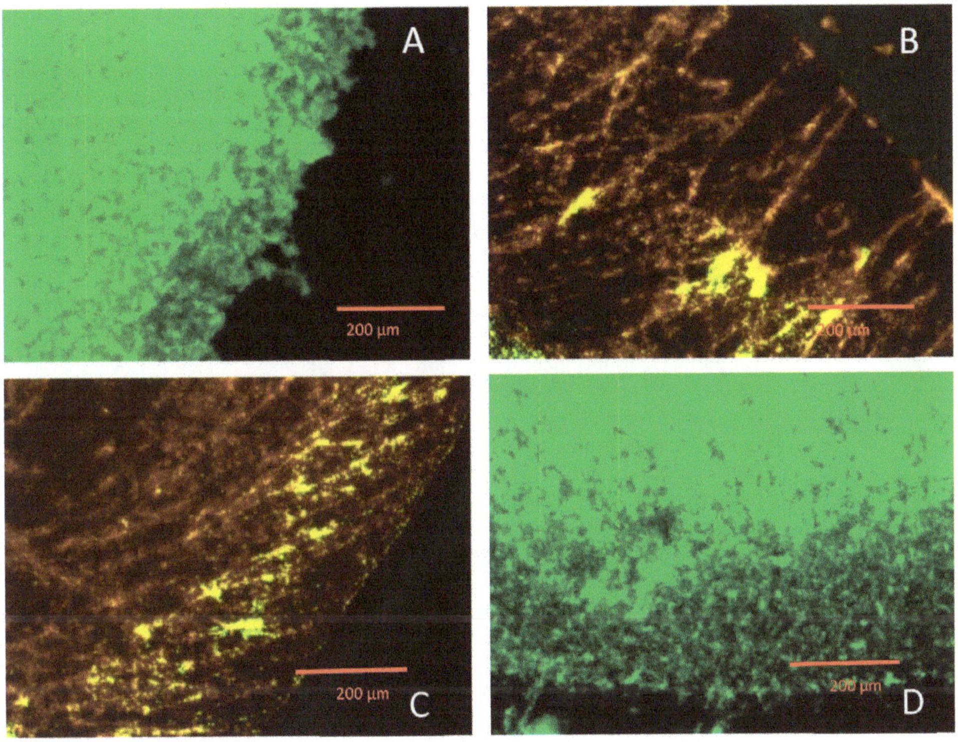

Figure 5. Stained microscopy images of *S. epidermidis* colonization after 20 h incubation, seeded at 1×10^7 CFU, on (**A**) the control ALDpp surface, (**B**) Magainin 2 graft coating, (**C**) Parasin 1 graft coating, and (**D**) on LL37 graft coating. The edges in the images are due to masks having been placed on samples.

The control ALDpp surface was completely overgrown after 20 h (Figure 5A) by apparently live bacteria; there was no evidence of dead (red-stained) bacteria. This is consistent with expectations; this fast-growing strain of *S. epidermidis* can colonize unprotected surfaces rapidly and proceed to biofilm formation.

The Magainin 2 coating showed larger numbers of bacteria after 20 h compared with 5 h. Many of the attached bacteria appear to be dead, but there is evidence of colony formation (the yellow clumps in Figure 5B) upon bacterial aggregation. Similarly, the Parasin 1 coating shows (Figure 5C) increasing numbers of attached bacteria, the majority of which seem to be dead according to the staining, but there are live bacteria visible and again the formation of clumps of bacteria. In contrast, the LL37 coating was overgrown by live bacteria after 20 h (Figure 5D). As Figure 4B shows a significant number of attached live bacteria after 4 h on the LL37 surface, it is not surprising that these attached live bacteria proceed to denser coverage. It appears from these data that grafted LL37 is relatively inefficient at killing attaching bacteria.

Whilst clear results were obtained with S. epidermidis, bacterial testing with *S. aureus* and *E. coli* was more difficult because these bacteria were far less inclined to colonize surfaces, even the ALDpp, resulting in very little colonization of the AMP surfaces over the first 5 h (data not shown). After 24 h, however, some colonization was evident on all samples, as for *S. epidermidis*. Lower amounts of bacteria resulted in less colonization up to 10 h.

These coatings were also tested in the presence of serum to determine whether serum proteins might affect or block the antibacterial activity, for example by adsorbing in a layer thick enough to "bury" the AMPs underneath them. For all coatings, however, the activity was not affected within the first 5 h; after 6 h, there were indications again that the surfaces were increasingly becoming colonised. It is difficult to separate any effect due to proteins from the increasing colonisation that also occurs in the absence of proteins, as shown above. All that can be concluded is that serum proteins do not immediately block the activity of the grafted AMPs and thus the coatings would be suitable for blood-contacting applications.

Samples were also soaked for 12 days and tested for activity. No reduction of activity was found. This is consistent with the known stability of interfacial amine bonds. It also verifies covalent grafting; if simply adsorbed on the surface (i.e., without formation of a covalent bond), these peptides, being soluble in PBS, should desorb from the surface.

3.3. Fibroblast Attachment

Many studies have reported that AMPs can be cytotoxic. With AMPs such as melittin, which is isolated from bee venom, it is an obvious concern. Studies with Magainin and Parasin have found them to be cytotoxic [37–40]. Magainin 2 was developed to limit the toxicity relative to its parent compound [39]. It is, however, not clear whether toxicity is still a relevant concern when peptides are covalently tethered, because most toxicity issues are manifested in renal and hepatic sites through the processes of breaking down the compounds and excreting the products. However, as these AMPs disrupt bacterial membranes, it is essential to study possible adverse effects on mammalian cells even when the peptides are surface-grafted.

Cell attachment was assessed on AMP-grafted surfaces and on two control surfaces (TCPS and ALDpp). Figure 6 shows data collected with primary human fibroblasts; with the fibroblast cell lines the data were closely analogous (not shown). The ALDpp surface gave results identical within experimental uncertainty to the attachment observed on the standard tissue culture surface TCPS. After grafting the AMPs onto the ALDpp layer, there was no difference in cell attachment and growth for LL37 and Magainin 2, whereas for Parasin 1 there were fewer cells on the surface and they showed some morphological abnormalities. However, these tests were done with a Parasin 1 sample that was of only 76% purity and it was impossible to elucidate what the cause of the toxicity might be. Subsequent tests using a Parasin 1 lot of 98.2% purity (grafted again on ALDpp) showed little to no toxicity and the cells were perhaps somewhat smaller, yet almost as numerous as on the ALDpp control and the other AMP-grafted samples. Accordingly, the AMP-grafted coatings look promising in that there appears to be no substantial cytotoxicity, though with Parasin 1 there should be further work before moving to an animal model, particularly assessing the purity of the peptide sample.

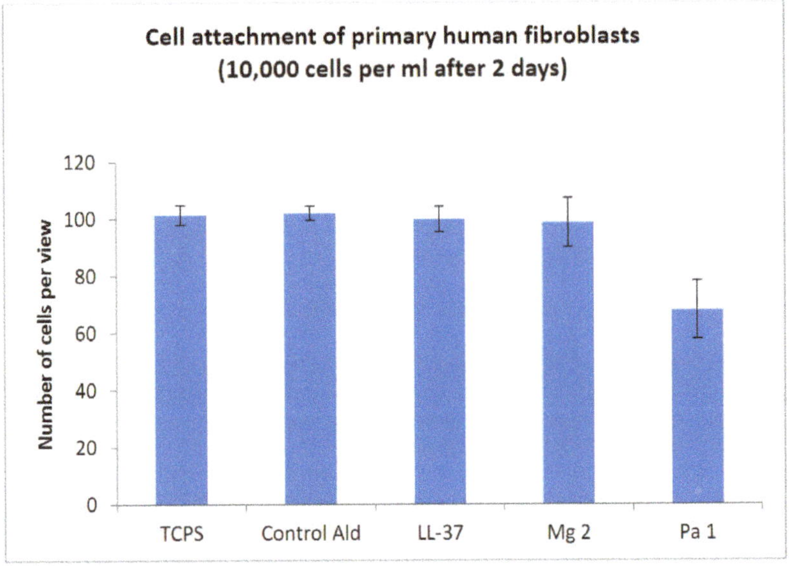

Figure 6. Attachment of primary human fibroblasts on control samples and on AMP-grafted surfaces (n = 3), counted per field of view at 10× magnification.

4. Discussion

AMPs have attracted considerable interest for the design of antibacterial coatings [13–15,22–24], but some aspects, particularly longer-term performance, are still in need of further study, as is the question of whether some AMPs might be less effective due to conformational changes or accessibility after grafting with specific immobilization chemistries. Their potential use as covalently grafted protective coatings needs to be informed by considerations such as mechanism(s) of action and possible cytotoxicity. AMPs exist in many tissues and various cells in a wide variety of plant and animal species. LL37 (active sequence 134–170) is cleaved extracellularly from hCAP18 by proteinase 3 when hCAP-18 is stimulated [41]. with the name LL37 denoting 37 amino acids starting with two leucines. It is a cationic, amphipathic α-helical peptide with broad-spectrum antimicrobial activity [42]. It is an effector of the innate immune system and expressed in leukocytes and epithelial cells, and in neutrophils and keratinocytes of inflamed skin. Magainin 2 also is a linear cationic α-helical peptide and the mechanism of activity is thought to be similar, through transmembrane pore formation, whilst Parasin 1, from catfish skin mucosa, is a histone H2A-derived amphipathic α-helical peptide upregulated via matrix metalloproteinase 2. It has pore-forming ability; the N-terminal binds to the membrane, the α-helical structure inserts either in a barrel-stave or snorkel manner, causing permeabilization. A single lysine residue near the N-terminal in the random coil region is essential to the mechanism of action; one possible explanation is that this anchors the peptide into the membrane, causing destabilization of the membrane and allowing the peptide to become embedded into the bacterial membrane. However, because of its similarity to the histone H2A-derived peptide buforin II, it is also believed to induce intracellular killing by binding to nucleic acids [43,44].

Such information is essential for selecting candidate AMPs and for rationalizing their performance as grafted surface coatings. The membrane destabilization mechanism of cationic amphipathic α-helical peptides could reasonably be expected to be maintained upon grafting to a solid surface; this is borne out by the observed high activity after 5 h of incubation. For Parasin 1 grafted onto ALDpp the putative intracellular activity would, of course, not be available upon grafting [43,44]. The observed activity indicates, however, that its ability for membrane permeabilization is retained upon surface grafting.

Another important consideration is to ascertain that the peptides have indeed been applied onto solid polymer surfaces to sufficient coverage, that adventitious other molecules, in particular organosilicones which can be found in proteins due to manufacturing processes or storage containers, are not present, and that the intended formation of covalent interfacial bonds has indeed taken place. Physico-chemical surface analysis methodologies are thus essential to ascertain that coatings are properly characterized prior to biological tests, thereby ensuring that interpretation of biological responses is not affected by artefacts and contaminants [35].

Extended soaking/washing followed by identical surface analysis experiments are essential for verifying that the coated peptides are indeed covalently grafted and thus are not able to detach during biological tests. Many reports on antimicrobial coatings were not supported with appropriate surface analysis and washing experiments, raising the question whether they were indeed fully covalently immobilised, or whether dissolving antibiotics might have affected biological testing [45]. Are some promising results possibly due to unrecognised diffusion of antibiotics intercepting bacteria approaching biomaterials surfaces? Our XPS and ToF-SIMS spectra clearly show grafting of all three peptides with good surface coverage; the precise coverage is difficult to determine due to the assumptions that have to be made when converting XPS atomic percentages to surface coverage, but our data are consistent with grafting densities of ~1/3 to 2/3 of a monolayer of peptides. Repeat surface analyses after extended soaking gave the same data and thus confirmed the covalent nature of the surface binding. This is not surprising; with their solubility in PBS these peptides would not be expected to show high affinity for (non-covalent) physisorption onto the relatively hydrophilic surface of the ALDpp and upon washing with PBS any small adsorbed amounts should readily dissolve off the surface.

Next, it is essential to check that an antimicrobial coating does not exhibit any significant cytotoxicity to mammalian cells. Peptides that destabilize bacterial membranes could also cause adverse effects to human cell membranes. With human-derived peptides this is less of a concern, but others need to be tested and there exists considerable information on this. For the present case, compared with Magainin 1, Magainin 2 has a lower hydrophobicity and was found to be non-haemolytic and non-toxic to human cells [39]. But it is conceivable that conformational changes upon surface grafting might alter the interactions of a grafted AMP with human cell membranes compared with those of a molecule in solution; the graft coating might be more cytotoxic or less cytotoxic. Moreover, this is likely to be dependent also on the grafting chemistry employed. Our results show no measurable toxicity for these three AMPs when grafted onto ALDpp.

Clearly, these grafted AMPs exhibit substantial antibacterial activity, consistent with previous reports on AMP coatings grafted using other immobilization chemistries. Yet, our data recorded after 20 h incubation also reveal a decrease in effectiveness upon continuing bacterial challenge. On the Magainin 2 and Parasin 1 graft coatings there is evidence of increasing numbers of attached dead bacteria and of initial formation of colonies. The LL37 graft coating was overgrown.

Many publications mention proteolytic degradation of AMPs [17,46], but while this may apply to a wound environment for example, it is doubtful whether proteolytic enzymes are at work in the test system used. A more likely explanation seems to be that on the LL37 coating some bacteria are not killed and thus end up initiating colonies. On the other two coatings such colony formation also seems to occur, to a much smaller extent. A putative interpretation is that grafted LL37 is relatively less efficient, or less rapid, at killing bacteria that come into contact with its surface, and eventually sufficient numbers survive.

The increasing presence with time of dead bacteria, attached with sufficient strength to resist rinsing, on the surface of the graft coatings poses concerns as to its implications if such coatings were to proceed to practical usage on biomedical devices. First, dead bacteria might eventually cover the grafted AMP layer and bacteria attaching on top of this layer of dead bacteria are not exposed to the action of surface-immobilised AMPs. This might be part of what happens on our LL37 coating. Secondly, the membrane permeabilization of attached bacteria might release endotoxins. This was beyond the scope of the present work.

Whilst the Magainin 2 and the Parasin 1 coatings showed high effectiveness in resisting biofilm formation, the presence of a significant number of dead bacteria raises important questions. It invites comparison with coatings that resist bacterial attachment by physico-chemical means such as hydration (non-fouling hydrogel coatings) [47–50], for which there have been no reports of increasing numbers of attached dead bacteria. Perhaps a coating that resists bacterial attachment altogether, as opposed to killing attaching bacteria, might be preferable for clinical applications on biomedical devices used in human medicine.

5. Conclusions

Our results show that AMPs immobilized via reductive amination onto a solid surface bearing aldehyde groups are tethered in a way that allows them to maintain an active conformation. A plasma polymer layer deposited from propanal has been found to be well suited as an adhesive interlayer for the grafting of the three AMPs. Plasma polymerization is a coating technology used to modify surfaces in a number of industries, and will enable transfer of the current grafting approach to a wide range of substrate materials. Detailed surface analyses showed that the intended grafting had indeed taken place and uniform graft coatings had been produced. The coatings did not exhibit significant cytotoxicity to primary human fibroblasts. All three AMPs were found to retain antibacterial activity when covalently grafted, with substantial reductions in bacterial colonization compared to the control plasma polymer surface. These AMPs on ALDpp can thus be used to make shorter-term effective antibacterial surface coatings for biomaterials by killing most attaching bacteria, but the detection of increasing numbers of attached dead bacteria, and some live bacteria, over time raises questions in regard to longer-term performance, and invites comparison with non-fouling coatings, which resist the attachment of dead (and live) bacteria and thereby avoid possible detrimental consequences arising from surface-bound dead bacteria.

Supplementary Materials: The following are available online at https://www.mdpi.com/2079-6412/11/1/68/s1.

Author Contributions: Conceptualization, S.S.G. and H.J.G.; methodology, S.S.G.; software, M.J.; formal analysis, S.S.G. and M.J.; investigation, S.S.G. and M.J.; resources, H.J.G.; data curation, S.S.G. and M.J.; writing—original draft preparation, S.S.G.; writing—review and editing, K.V. and H.J.G.; supervision, K.V. and H.J.G.; project administration, K.V. and H.J.G.; funding acquisition, H.J.G. All authors have read and agreed to the published version of the manuscript.

Funding: This work was supported by the CRC for Wound Management Innovation, which is acknowledged for granting a PhD scholarship to SSG. KV thanks the NHMRC for Fellowship APP1122825 and Project grant APP1032738.

Institutional Review Board Statement: Not applicable.

Informed Consent Statement: Not applicable.

Data Availability Statement: Data are contained within the article and the supplementary material.

Acknowledgments: We gratefully acknowledge subsidised hands-on access to the surface analytical instruments XPS and ToF-SIMS at the UniSA node of Microscopy Australia, a national network funded under the Australian Government's NCRIS scheme.

Conflicts of Interest: The authors declare no conflict of interest. The funders had no role in the design of the study; in the collection, analyses, or interpretation of data; in the writing of the manuscript, or in the decision to publish the results.

References

1. Perez-Koehler, B.; Bayon, Y.; Bellón, J.M. Mesh infection and hernia repair: A review. *Surg. Infect.* **2016**, *17*, 124–137. [CrossRef] [PubMed]
2. Rattanawong, P.; Kewcharoen, J.; Mekraksakit, P.; Mekritthikrai, R.; Prasitlumkum, N.; Vutthikraivit, W.; Putthapiban, P.; Dworkin, J. Device infections in implantable cardioverter defibrillators versus permanent pacemakers: A systematic review and meta-analysis. *J. Cardiovasc. Electrophysiol.* **2019**, *30*, 1053–1065. [CrossRef] [PubMed]
3. Scotland, K.B.; Lo, J.; Grgic, T.; Lange, D. Ureteral stent-associated infection and sepsis: Pathogenesis and prevention: A review. *Biofouling* **2019**, *35*, 117–127. [CrossRef] [PubMed]
4. Veerachamy, S.; Yarlagadda, T.; Manivasagam, G.; Yarlagadda, P.K. Bacterial adherence and biofilm formation on medical implants: A review. *Proc. Inst. Mech. Eng. Part H J. Eng. Med.* **2014**, *228*, 1083–1099. [CrossRef]
5. Vickery, K.; Hu, H.; Jacombs, A.S.; Bradshaw, D.A.; Deva, A.K. A review of bacterial biofilms and their role in device-associated infection. *Healthc. Infect.* **2013**, *18*, 61–66. [CrossRef]
6. Vasilev, K.; Cook, J.; Griesser, H.J. Antibacterial surfaces for biomedical devices. *Expert Rev. Med. Devices* **2009**, *6*, 553–567. [CrossRef]
7. Vasilev, K.; Griesser, S.S.; Griesser, H.J. Antibacterial surfaces and coatings produced by plasma techniques. *Plasma Process. Polym.* **2011**, *8*, 1010–1023. [CrossRef]
8. Campoccia, D.; Montanaro, L.; Arciola, C.R. A review of the biomaterials technologies for infection-resistant surfaces. *Biomaterials* **2013**, *34*, 8533–8554.
9. Chouirfa, H.; Bouloussa, H.; Migonney, V.; Falentin-Daudré, C. Review of titanium surface modification techniques and coatings for antibacterial applications. *Acta Biomater.* **2019**, *83*, 37–54. [CrossRef]
10. Baltzer, S.A.; Brown, M.H. Antimicrobial peptides–promising alternatives to conventional antibiotics. *J. Mol. Microbiol. Biotechnol.* **2011**, *20*, 228–235.
11. Brogden, K.A. Antimicrobial peptides: Pore formers or metabolic inhibitors in bacteria? *Nat. Rev. Microbiol.* **2005**, *3*, 238–250. [PubMed]
12. Yasir, M.; Willcox, M.D.P.; Dutta, D. Action of antimicrobial peptides against bacterial biofilms. *Materials* **2018**, *11*, 2468.
13. Magana, M.; Pushpanathan, M.; Santos, A.L.; Leanse, L.; Fernandez, M.; Ioannidis, A.; Giulianotti, M.A.; Apidianakis, Y.; Bradfute, S.; Ferguson, A.L. The value of antimicrobial peptides in the age of resistance. *Lancet Infect. Dis.* **2020**, *20*, e216–e230. [PubMed]
14. Haney, E.F.; Straus, S.K.; Hancock, R.E. Reassessing the host defense peptide landscape. *Front. Chem.* **2019**, *7*, 43. [PubMed]
15. Ting, D.S.J.; Beuerman, R.W.; Dua, H.S.; Lakshminarayanan, R.; Mohammed, I. Strategies in Translating the Therapeutic Potentials of Host Defense Peptides. *Front. Immunol.* **2020**, *11*, 983. [CrossRef] [PubMed]
16. Barns, K.J.; Weisshaar, J.C. Real-time attack of LL-37 on single Bacillus subtilis cells. *Biochim. Biophys. Acta (BBA) Biomembr.* **2013**, *1828*, 1511–1520. [CrossRef] [PubMed]
17. Costa, F.; Carvalho, I.F.; Montelaro, R.C.; Gomes, P.; Martins, M.C.L. Covalent immobilization of antimicrobial peptides (AMPs) onto biomaterial surfaces. *Acta Biomater.* **2011**, *7*, 1431–1440.
18. Wimley, W.C.; Hristova, K. Antimicrobial peptides: Successes, challenges and unanswered questions. *J. Membr. Biol.* **2011**, *239*, 27–34.
19. Yasir, M.; Dutta, D.; Hossain, K.R.; Chen, R.; Ho, K.K.; Kuppusamy, R.; Clarke, R.J.; Kumar, N.; Willcox, M.D. Mechanism of action of surface immobilized antimicrobial peptides against Pseudomonas aeruginosa. *Front. Microbiol.* **2020**, *10*, 3053. [CrossRef]
20. Yasir, M.; Dutta, D.; Kumar, N.; Willcox, M.D. Interaction of the surface bound antimicrobial peptides melimine and Mel4 with Staphylococcus aureus. *Biofouling* **2020**. [CrossRef]
21. Yazici, H.; O'Neill, M.B.; Kacar, T.; Wilson, B.R.; Oren, E.E.; Sarikaya, M.; Tamerler, C. Engineered chimeric peptides as antimicrobial surface coating agents toward infection-free implants. *ACS Appl. Mater. Interfaces* **2016**, *8*, 5070–5081. [CrossRef] [PubMed]
22. Riool, M.; de Breij, A.; Drijfhout, J.W.; Nibbering, P.H.; Zaat, S.A. Antimicrobial peptides in biomedical device manufacturing. *Front. Chem.* **2017**, *5*, 63. [CrossRef]
23. Pinto, I.B.; dos Santos Machado, L.; Meneguetti, B.T.; Nogueira, M.L.; Carvalho, C.M.E.; Roel, A.R.; Franco, O.L. Utilization of antimicrobial peptides, analogues and mimics in creating antimicrobial surfaces and bio-materials. *Biochem. Eng. J.* **2019**, *150*, 107237. [CrossRef]
24. Kazemzadeh-Narbat, M.; Cheng, H.; Chabok, R.; Alvarez, M.M.; De La Fuente-Nunez, C.; Phillips, K.S.; Khademhosseini, A. Strategies for antimicrobial peptide coatings on medical devices: A review and regulatory science perspective. *Crit. Rev. Biotechnol.* **2020**, *41*, 1–27. [CrossRef] [PubMed]
25. Siow, K.S.; Britcher, L.; Kumar, S.; Griesser, H.J. Plasma methods for the generation of chemically reactive surfaces for biomolecule immobilization and cell colonization—A review. *Plasma Process. Polym.* **2006**, *3*, 392–418.
26. Vasilev, K.; Michelmore, A.; Griesser, H.J.; Short, R.D. Substrate influence on the initial growth phase of plasma-deposited polymer films. *Chem. Commun.* **2009**, *24*, 3600–3602. [CrossRef] [PubMed]
27. Vasilev, K.; Michelmore, A.; Martinek, P.; Chan, J.; Sah, V.; Griesser, H.J.; Short, R.D. Early stages of growth of plasma polymer coatings deposited from nitrogen-and oxygen-containing monomers. *Plasma Process. Polym.* **2010**, *7*, 824–835. [CrossRef]
28. Hernandez-Lopez, J.; Bauer, R.; Chang, W.-S.; Glasser, G.; Grebel-Koehler, D.; Klapper, M.; Kreiter, M.; Leclaire, J.; Majoral, J.-P.; Mittler, S. Functional polymers as nanoscopic building blocks. *Mater. Sci. Eng. C* **2003**, *23*, 267–274. [CrossRef]

29. Griesser, H.J. Small scale reactor for plasma processing of moving substrate web. *Vacuum* **1989**, *39*, 485–488. [CrossRef]
30. Coad, B.R.; Scholz, T.; Vasilev, K.; Hayball, J.D.; Short, R.D.; Griesser, H.J. Functionality of proteins bound to plasma polymer surfaces. *ACS Appl. Mater. Interfaces* **2012**, *4*, 2455–2463. [CrossRef]
31. Briggs, D. *Surface Analysis of Polymers by XPS and Static SIMS*; Cambridge University Press: Cambridge, UK, 1998.
32. Jasieniak, M.; Graham, D.; Kingshott, P.; Gamble, L.; Griesser, H.J. Surface Analysis of Biomaterials. In *Handbook of Surface and Interface Analysis*, 2nd ed.; Riviere, J.P., Myhra, S., Eds.; CRC Press: Boca Raton, FL, USA, 2009; pp. 529–564.
33. MacNeil, S.; Shepherd, J.; Smith, L. Production of tissue-engineered skin and oral mucosa for clinical and experimental use. In *3D Cell Culture*; Springer: Berlin/Heidelberg, Germany, 2011; pp. 129–153.
34. Coad, B.R.; Vasilev, K.; Diener, K.R.; Hayball, J.D.; Short, R.D.; Griesser, H.J. Immobilized streptavidin gradients as bioconjugation platforms. *Langmuir* **2012**, *28*, 2710–2717. [PubMed]
35. Castner, D.G.; Ratner, B.D. Biomedical surface science: Foundations to frontiers. *Surface Sci.* **2002**, *500*, 28–60. [CrossRef]
36. Wagner, M.S.; Castner, D.G. Characterization of Adsorbed Protein Films by Time-of-Flight Secondary Ion Mass Spectrometry with Principal Component Analysis. *Langmuir* **2001**, *17*, 4649–4660. [CrossRef]
37. Jang, S.A.; Kim, H.; Lee, J.Y.; Shin, J.R.; Kim, D.J.; Cho, J.H.; Kim, S.C. Mechanism of action and specificity of antimicrobial peptides designed based on buforin IIb. *Peptides* **2012**, *34*, 283–289. [CrossRef] [PubMed]
38. Nascimento, J.M.; Franco, O.L.; Oliveira, M.D.L.; Andrade, C.A.S. Evaluation of magainin I interactions with lipid membranes: An optical and electrochemical study. *Chem. Phys. Lipids* **2012**, *165*, 537–544. [CrossRef] [PubMed]
39. Tachi, T.; Epand, R.F.; Epand, R.M.; Matsuzaki, K. Position-dependent hydrophobicity of the antimicrobial magainin peptide affects the mode of peptide-lipid interactions and selective toxicity. *Biochemistry* **2002**, *41*, 10723–10731. [CrossRef]
40. Vila-Farres, X.; de la Maria, C.G.; Lopez-Rojas, R.; Pachon, J.; Giralt, E.; Vila, J. In vitro activity of several antimicrobial peptides against colistin-susceptible and colistin-resistant *Acinetobacter baumannii*. *Clin. Microbiol. Infect.* **2012**, *18*, 383–387.
41. Sørensen, O.E.; Follin, P.; Johnsen, A.H.; Calafat, J.; Tjabringa, G.S.; Hiemstra, P.S.; Borregaard, N. Human cathelicidin, hCAP-18, is processed to the antimicrobial peptide LL-37 by extracellular cleavage with proteinase 3. *Blood J. Am. Soc. Hematol.* **2001**, *97*, 3951–3959. [CrossRef]
42. Dürr, U.H.; Sudheendra, U.; Ramamoorthy, A. LL-37, the only human member of the cathelicidin family of antimicrobial peptides. *Biochim. Biophys. Acta (BBA) Biomembr.* **2006**, *1758*, 1408–1425. [CrossRef]
43. Koo, Y.S.; Kim, J.M.; Park, I.Y.; Yu, B.J.; Jang, S.A.; Kim, K.-S.; Park, C.B.; Cho, J.H.; Kim, S.C. Structure–activity relations of parasin I, a histone H2A-derived antimicrobial peptide. *Peptides* **2008**, *29*, 1102–1108. [CrossRef]
44. Uyterhoeven, E.T.; Butler, C.H.; Ko, D.; Elmore, D.E. Investigating the nucleic acid interactions and antimicrobial mechanism of buforin II. *FEBS Lett.* **2008**, *582*, 1715–1718. [CrossRef] [PubMed]
45. Naderi, J.; Giles, C.; Saboohi, S.; Griesser, H.J.; Coad, B.R. Surface-grafted antimicrobial drugs: Possible misinterpretation of mechanism of action. *Biointerphases* **2018**, *13*, 06E409. [CrossRef] [PubMed]
46. Mowery, B.P.; Lee, S.E.; Kissounko, D.A.; Epand, R.F.; Epand, R.M.; Weisblum, B.; Stahl, S.S.; Gellman, S.H. Mimicry of antimicrobial host-defense peptides by random copolymers. *J. Am. Chem. Soc.* **2007**, *129*, 15474–15476. [CrossRef] [PubMed]
47. Cavallaro, A.A.; Macgregor-Ramiasa, M.N.; Vasilev, K. Antibiofouling properties of plasma-deposited oxazoline-based thin films. *ACS Appl. Mater. Interfaces* **2016**, *8*, 6354–6362. [CrossRef] [PubMed]
48. Kingshott, P.; Wei, J.; Bagge-Ravn, D.; Gadegaard, N.; Gram, L. Covalent attachment of poly (ethylene glycol) to surfaces, critical for reducing bacterial adhesion. *Langmuir* **2003**, *19*, 6912–6921. [CrossRef]
49. Maddikeri, R.; Tosatti, S.; Schuler, M.; Chessari, S.; Textor, M.; Richards, R.; Harris, L. Reduced medical infection related bacterial strains adhesion on bioactive RGD modified titanium surfaces: A first step toward cell selective surfaces. *J. Biomed. Mater. Res. Part A* **2008**, *84*, 425–435. [CrossRef]
50. Sileika, T.S.; Kim, H.-D.; Maniak, P.; Messersmith, P.B. Antibacterial performance of polydopamine-modified polymer surfaces containing passive and active components. *ACS Appl. Mater. Interfaces* **2011**, *3*, 4602–4610. [CrossRef]

coatings

Article

Ag Functionalization of Al-Doped ZnO Nanostructured Coatings on PLA Substrate for Antibacterial Applications

Daniele Valerini [1,*], Loredana Tammaro [2], Giovanni Vigliotta [3], Enrica Picariello [3], Francesco Banfi [4], Emanuele Cavaliere [5], Luca Ciambriello [5] and Luca Gavioli [5]

1 Laboratory of Functional Materials and Technologies for Sustainable Applications (SSPT-PROMAS-MATAS), ENEA—Italian National Agency for New Technologies, Energy and Sustainable Economic Development, S.S. 7 Appia, km 706, 72100 Brindisi, Italy

2 Nanomaterials and Devices Laboratory (SSPT-PROMAS-NANO), ENEA—Italian National Agency for New Technologies, Energy and Sustainable Economic Development, Piazzale E. Fermi, 1, Portici, 80055 Napoli, Italy; loredana.tammaro@enea.it

3 Department of Chemistry and Biology "A. Zambelli", University of Salerno, Via Giovanni Paolo II, 112, Fisciano, 80144 Salerno, Italy; gvigliotta@unisa.it (G.V.); enrica.picariello@hotmail.it (E.P.)

4 FemtoNanoOptics group, Université de Lyon, CNRS, Université Claude Bernard Lyon 1, Institut Lumière Matière, F-69622 Villeurbanne, France; francesco.banfi@univ-lyon1.fr

5 Interdisciplinary Laboratories for Advanced Materials Physics (i-LAMP) and Dipartimento di Matematica e Fisica, Università Cattolica del Sacro Cuore, Via Musei 41, 25121 Brescia, Italy; Emanuele.Cavaliere@unicatt.it (E.C.); luca.ciambriello@unicatt.it (L.C.); luca.gavioli@unicatt.it (L.G.)

* Correspondence: daniele.valerini@enea.it

Received: 20 November 2020; Accepted: 14 December 2020; Published: 17 December 2020

Abstract: Developing smart, environmentally friendly, and effective antibacterial surfaces is fundamental to contrast the diffusion of human infections and diseases for applications in the biomedical and food packaging sectors. To this purpose, here we combine aluminum-doped zinc oxide (AZO) and Ag to grow nanostructured composite coatings on bioplastic polylactide (PLA) substrates. The AZO layers are grown by RF magnetron sputtering, and then functionalized with Ag in atomic form by RF magnetron sputtering and in form of nanoparticles by supersonic cluster beam deposition. We compare the morphology, wettability, and antimicrobial performance of the nanostructured coatings obtained by the two methods. The different growth modes in the two techniques used for Ag functionalization are found to produce some differences in the surface morphology, which, however, do not induce significant differences in the wettability and antimicrobial response of the coatings. The antibacterial activity is investigated against *Escherichia coli* and *Staphylococcus aureus* as representatives of Gram-negative and Gram-positive bacteria, respectively. A preferential antimicrobial action of Ag on the first species and of AZO on the second one is evidenced. Through their combination, we obtain a hybrid composite coating taking advantage of the synergistic dual action of the two materials deposited, with a total bacterial suppression within few minutes for the first species and few hours for the second one, thus representing a valuable solution as a wide-spectrum bactericidal device.

Keywords: antimicrobial coatings; aluminum-doped zinc oxide (AZO); RF sputtering (RFS); supersonic cluster beam deposition (SCBD); silver nanoparticles; atomic force microscopy (AFM); health; biomedical applications; food packaging

1. Introduction

High-performance functional coatings represent one of the most effective strategy to confer antimicrobial properties to surfaces where avoiding human infections and disease diffusion are primary issues [1]. Such surfaces should be able to kill pathogen microorganisms or inhibit their growth and biofilm formation, as well as to prevent diffusion of foodborne diseases and reduce food degradation to preserve quality and increase shelf-life [2]. The use of antimicrobial surfaces can limit the cross-contamination phenomena due to the transfer of microorganisms from a contaminated surface to another. Moreover, due to the multiple simultaneous mechanisms of actions introduced by the coating materials, bacterial resistance is harder to be developed with respect to conventional antibiotics [3]. The development of efficient antibacterial materials is therefore a key aspect for a wide range of applications—such as biomedical devices (medical equipment, surgery tools, implants, etc.), food and beverage packaging materials, and textiles—just to mention a few. On the other hand, employed materials should also ensure environmental safety, no secondary toxicity to human health, or even biocompatibility, depending on the final use. To this aim, proper design of composition and structure of surface coatings turns out to be essential to achieve all the required properties. Apart from intrinsic biocide action of the coating material itself, other characteristics may be taken into account to enhance the antimicrobial efficacy, such as tuning of surface features (nanostructuration, roughness, etc.) and suitable combination of multiple active elements.

ZnO-based materials provide a strong bactericide action and physico-chemical stability, although the safety for human health is still debated in literature, especially for nanoparticles (NPs) [4,5]. Doping or combining ZnO with other elements can enhance its antimicrobial properties and favorably modify other characteristics like mechanical, optical, and gas/water vapor barrier properties. For example, Al-doped ZnO nanomaterials have been reported as effective antibacterial agents against different Gram-positive and Gram-negative species [6–8]. In the first report dealing with antimicrobial properties of AZO coatings [6], nanostructured Al-doped ZnO films were grown by sputtering a target source composed of mixed ZnO and Al_2O_3, since both components exhibit antibacterial properties [4,5] and, at the same time, they are both considered biocompatible and authorized as additives in food contact plastic materials [8]. Aiming to develop an environmentally friendly and health-safe composite, in our experiments the coatings were applied onto polylactide, a nature-derived polymer with several useful properties, biocompatible, bioresorbable, biodegradable and compostable [6,8]. In our previous works, such AZO coatings on PLA presented a polycrystalline phase composition, they demonstrated long-term stability (low release of material in physiological saline solution) and were shown to possess strong antibacterial action against different Gram-positive and Gram-negative species, without any significant toxicity detected on some human cell lines tested. To further improve the antimicrobial performance of those coatings, in the present work we add silver as supplementary material well-known for its antibacterial properties [9]. Both Ag and ZnO-based materials are identified to act against microbial species by means of several mechanisms, like release of active species (e.g., metal ions and reactive oxygen species ROS) or release of particles, with consequent cell membrane rupture, particle internalization inducing metabolic alterations, electrostatic interactions between the bacterial wall and the active materials and mechanical damaging of cell membrane and structure [4–9]. As a consequence, their arrangement into composite coatings can be expected to enhance the total antimicrobial action through the combined action of the two materials. In particular, herein two different methods are explored to functionalize the AZO surface with the Ag add-on: RF magnetron sputtering (RFS) and supersonic cluster beam deposition (SCBD). Differently from the sputtering deposition, where the depositing species are mainly constituted by atoms and ions, in the SCBD technique Ag NPs are already formed in the gas phase and then directed towards the substrate [10]. As a consequence, the diverse growth regimes occurring in the two methods may give rise to different surface morphologies of the Ag-functionalized AZO deposits on PLA. Atomic force microscopy, contact angle measurements and antibacterial tests are then conducted to investigate the differences in the properties of the mixed coatings and to compare the responses of the single layer coatings (only AZO on PLA and only Ag on PLA) with those of the multicomponent hybrid layers (Ag-functionalized AZO coatings on PLA).

2. Materials and Methods

AZO coatings were deposited at room temperature by RF magnetron sputtering in a vacuum chamber (Sistec thin film equipment, Angelantoni Group, Italy) pumped down to a base pressure of 1×10^{-4} Pa. A $ZnO:Al_2O_3$ target (composition 98:2 wt.%, purity 99.999%, diameter 10 cm) was sputtered at RF power of 150 W in Ar + O_2 atmosphere at process pressure of 3 Pa, with relative flux percentage of 90% Ar and 10% O_2. Prior to the deposition, the cathode power was gradually increased until the set point of 150 W in about 30 min in the same Ar + O_2 atmosphere, and further cleaning and conditioning of the target surface was carried on for additional 10 min before to start the deposition process on the substrates. The coatings were simultaneously deposited on PLA films (extruded as previously reported [6]) and reference silicon substrates, passing under the sputtering target through rotation at 5 rpm for 20 min, resulting in AZO thickness of 30 nm, as measured by stylus profilometry.

The AZO layers were functionalized by depositing Ag under different growth regimes through the use of two different techniques: RF magnetron sputtering and supersonic cluster beam deposition. For the Ag-functionalization by means of RFS, a set of AZO samples was coated in the same equipment described above, by sputtering Ag at 50 W in Ar atmosphere, with the substrates passing under the target for a fixed number of 11 cycles at 5 rpm. Another set of samples was functionalized by SCBD at room temperature and at a base pressure of 1×10^{-4} Pa, following the same procedure described elsewhere [11–13], depositing Ag NPs on the sputter-deposited AZO films. In brief, SCBD relies on pulsed plasma ablation of a 99.99% purity Ag rod (ACI alloys), followed by NP condensation and expansion through a custom designed aerodynamic focusing system, directing the NPs to the substrate. The 6-nm thickness of the Ag layers was measured by atomic force microscopy on a silicon reference substrate deposited together with the PLA substrates. In addition to the Ag-functionalized AZO coatings (labeled as Ag/AZO), also single AZO and Ag coatings on PLA were deposited for comparison. For the discussion of the results, hereafter the samples deposited on PLA substrates will be labeled as AZO, Ag(RFS), Ag(SCBD), Ag(RFS)/AZO, and Ag(SCBD)/AZO, while the reference Ag samples deposited on silicon substrates will be labeled as Ag(RFS)/Si and Ag(SCBD)/Si, where the acronym 'RFS' or 'SCBD' in brackets stands for the technique used for the Ag deposition.

Atomic force microscopy data were obtained by an AFM (Park NX10, Park Systems, Suwon, Korea) in non-contact and tapping mode. PPP-NHCR tips (resonance frequency in the 250–280 kHz range, nominal tip radius 10 nm) were employed. The raw AFM data were analyzed with Gwyddion, extracting the grain size by a watershed algorithm.

Contact angles of ultrapure water and diiodomethane on coatings deposited on PLA substrates were measured by sessile drop method using an OCA 20 (Dataphysics, Filderstadt, Germany) goniometer, and data were collected with SCA 202 software (version 3.4.3 build 76). Equilibrium (static) contact angles were measured for 1 μL droplet volumes. Measurements were made on 10 different locations for each condition and the average value was reported with its standard deviation. To compute the surface free energy from measured contact angles the Owens, Wendt, Rabel, and Kaelble (OWRK) method [14] was applied and the related polar and dispersive components were evaluated by the Ström et al. equation [15], where the dispersion (non-polar) and polar components of the surface tension of the liquids employed were 21.8 and 51.0 mN/m for water and 50.8 and 0 mN/m for diiodomethane, respectively.

Effect of the different coated films on microorganism survival was evaluated as previously reported [8], with few variations. Briefly, *Escherichia coli* and *Staphylococcus aureus* were pre-inoculated aerobically for 12 h at 37 °C in Luria-Bertani (LB) medium (10 $g \cdot L^{-1}$ tryptone, 5 $g \cdot L^{-1}$ yeast extract, 10 $g \cdot L^{-1}$ NaCl), with constant shaking at 250 rpm. Bacteria were collected by centrifugation for 10 min at 3500 g, re-suspended at cellular density of 0.01 and 0.02 OD600 for *E. coli* and *S. aureus*, respectively (about 1 and 3×10^7 colony forming units, CFUs), in the presence of 1 cm^2 of sample (each cut into four equal parts) and incubated at 37 °C under constant agitation at 50 rpm by vertical rotator. *E. coli* was re-suspended in sterile distilled water, while *S. aureus*, more sensitive to oligotrophic conditions was re-suspended in peptone water (1 g·L^{-1} peptone, 5 g·L^{-1} NaCl). A control with uncoated PLA was also inserted. At the times indicated 100 µl of each sample was suitably diluted, distributed on LB agar dishes (15 g·L^{-1} agar) and incubated for 18–24 h at 37 °C. Subsequently, the number of CFU/mL was quantified for each sample and the survival kinetics was determined as the percentage variation over time of CFUs with respect to the initial time t = 0. The values were plotted on a semilogarithmic scale as the mean ± standard deviation of three independent tests. The obtained CFUs were also used to calculate the antibacterial activity (A) [16]. For this purpose, the following formula was applied: $A = F - G$, where F represents the growth values in the presence of uncoated PLA, while G corresponds to the growth values with the coated samples. They are calculated according to the formula $F = (\text{Log } C_t - \text{Log } C_{t0})$ and $G = (\text{Log } T_t - \text{Log } T_{t0})$. C and T are CFUs detected for uncoated PLA control and coated samples, respectively, at different times up to 8 h. The PLA films were considered 'antimicrobial' when achieving an A greater than 2 (reduction in bacteria number >99%). Values were reported as the mean ± standard deviation of three independent analyses. The significance of the differences in the antimicrobial activity among the various treatments was evaluated by two-way ANOVAs (analyses of variance), followed by Tukey post hoc tests (for $\alpha = 0.05$).

3. Results and Discussion

3.1. Atomic Force Microscopy (AFM) Analyses

Three dimensional (3D) representation of the AFM data results of the AZO samples functionalized with Ag by the two different techniques, presented in Figure 1a,c, clearly show the presence of some wrinkling effects of the sample surface, corresponding to a rms roughness of 35 nm deriving from the AZO deposition on PLA. This rms value is in line with that one previously measured on AZO samples deposited at high RF power on PLA and ascribed to wrinkling effects induced on the polymer surface by the particle bombardment during the AZO sputtering deposition, as extensively described in our previous work [8]. To obtain the nanoscale grain size distribution it has been necessary to remove such wrinkling effect by a polynomial regression, and the resulting 3D images are reported in Figure 1b,d for Ag(RFS)/AZO and Ag(SCBD)/AZO, respectively.

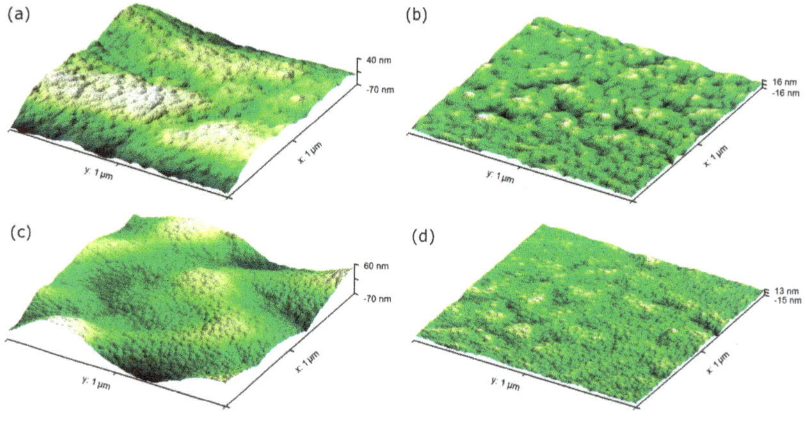

Figure 1. 3D AFM images of Ag/AZO coatings on PLA before (left) and after (right) fine correction: (**a**) and (**b**) Ag(RFS)/AZO; (**c**) and (**d**) Ag(SCBD)/AZO. For all images, the X:Y:Z scale ratio is 1:1:1.

After removing the background, the resulting AFM images of samples deposited on PLA substrates are shown in Figure 2 and the related rms roughness values are listed in Table 1.

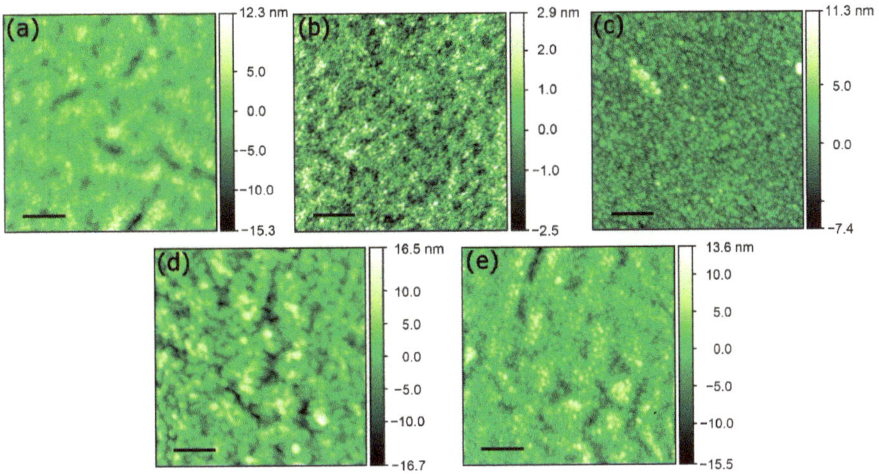

Figure 2. (1 × 1) μm^2 AFM images of films deposited on PLA: (**a**) AZO; (**b**) Ag(RFS); (**c**) Ag(SCDB); (**d**) Ag(RFS)/AZO; (**e**) Ag(SCBD)/AZO. Scale bar for all panels is 200 nm. Images in panels (**a**,**d**,**e**) have been subtracted a polynomial background to remove the roughness of the AZO/PLA layer.

Table 1. Rms roughness, height (mode of the distribution), and width (standard deviation) of height distributions obtained from the AFM analyses for coatings deposited on PLA and for reference Ag samples deposited on silicon.

Sample	Rms Roughness (±0.1 nm)	Height (±0.2 nm)	Distribution Width (±0.1 nm)
AZO	2.4	1.8	1.6
Ag(RFS)	0.9	1.2	1.1
Ag(SCBD)	1.9	2.2	1.9
Ag(RFS)/AZO	4.2	5.0	3.9
Ag(SCBD)/AZO	3.3	3.7	2.5
Ag(RFS)/Si	3.6	4.3	4.4
Ag(SCBD)/Si	1.1	1.2	1.4

The fine morphology appears to be constituted by nanometric clusters uniformly distributed over the sample surface and close one each other. In such samples, quantification of the NP size distribution from the projected surface area is affected by convolution effects of the tip. We hence obtain the grain size by measuring the height difference between the center of each grain and the height at the grain border, assuming that we are observing the top half of a NP with a spherical shape. Figure 3 displays the radius distribution of the different samples, that has been fitted with either Gaussian or lognormal distribution functions depending on the sample type. The RFS-synthesized samples are fitted with a Gaussian distribution reflecting the growth mode of the grains in an atom-by-atom mode. The SCBD-synthesized samples have been fitted with a lognormal distribution, since these layers are obtained by depositing directly NPs and not atoms, and hence reflect the typical size distribution of clusters produced by supersonic beams also for different materials [13,17–21]. The values of mode and width obtained from the fit of the height distributions are reported in Table 1.

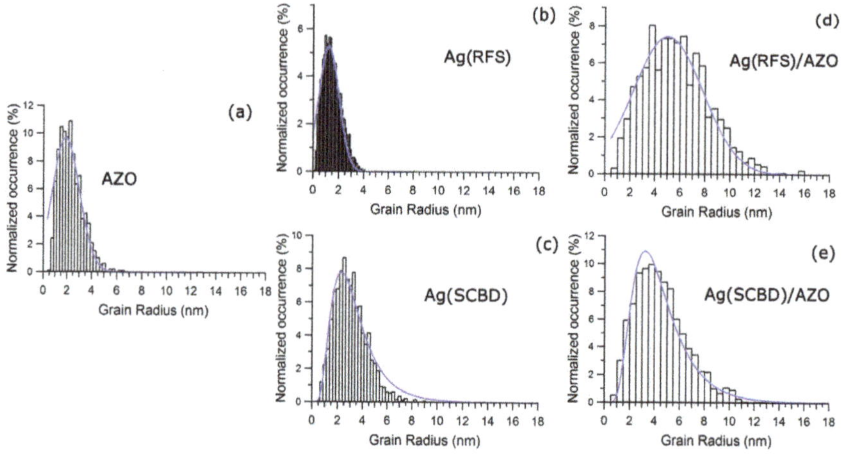

Figure 3. Nanoparticle height distributions obtained from the analysis of AFM data for samples deposited on PLA substrate: (**a**) AZO; (**b**) Ag(RFS); (**c**) Ag(SCDB); (**d**) Ag(RFS)/AZO; (**e**) Ag(SCBD)/AZO. The blue curves represent the best fit functions as described in the discussion.

The different growth mode of the Ag layers obtained from the two synthesis methods can be noticed already from the AFM images of samples directly deposited on PLA, shown in Figure 2b for Ag(RFS) and Figure 2c for Ag(SCBD) respectively, and from the striking difference of the related size distributions reported in Figure 3b,c. On the bare PLA, the mode of the Ag grain size is 1.2 ± 0.2 nm for the Ag(RFS) sample and 2.2 ± 0.2 nm for the Ag(SCBD) sample (see Table 1). On the AZO/PLA substrate the mode of the size distribution is 5.0 ± 0.2 nm for Ag deposited by RFS and 3.7 ± 0.2 nm for Ag deposited by SCBD. It is worth noting that for Ag deposited on reference silicon substrates (distributions not shown in Figure 3 for clarity, but fit values reported in Table 1), the mode of the distribution is 4.3 ± 0.2 nm for the Ag(RFS) and 1.2 ± 0.2 nm for the Ag(SCBD) sample. This behavior can be ascribed to the constituents produced by the two synthesis methods (mainly atoms and ions in RF sputtering, NPs in SCBD), to the substrate diffusion coefficient and to the kinetic energy of the atoms or NPs. The species synthesized by SCBD have an average energy of 0.1–0.5 eV/atom [21,22], while those produced by RF sputtering process have an energy of ~10 eV/atom [23,24], two orders of magnitude higher. On a silicon substrate, that is crystalline and with a rms lower than 0.2 nm, the landing energy of the RF sputtered atoms account for a large atomic diffusion over the surface, promoting the atoms coalescence into 4.3 ± 0.2 nm clusters. This behavior is almost absent for the less energetic SCBD NPs, which maintains their integrity upon landing, thus resulting in a mode of 1.2 ± 0.2 nm. On PLA, the atoms and the NPs may stick on the polymer surface or be easily implanted underneath the surface, thus hindering their mobility and blocking their coalescence into bigger clusters. This results in an almost four times reduction of the distribution mode (down to 1.2 ± 0.2 nm) for the Ag(RFS) deposit on PLA with respect to Si, while it is only slightly increasing the distribution mode (2.2 ± 0.2 nm) for the Ag(SCBD) sample. The single AZO coating on PLA forms grains of 1.8 ± 0.2 nm and presents a rms of 2.4 ± 0.1 nm, while these values increase in both the Ag-functionalized samples, as expected due to the superimposition of the Ag particles with the underlying AZO grains. In particular, the size distribution mode in the Ag(SCBD)/AZO sample increases to 3.7 ± 0.2 nm, mainly due to such cluster superimposition, since the NPs deposited by this method are not modified upon landing, as mentioned above. Differently, the observed grain size on the Ag(RFS)/AZO present a further increase, up to 5.0 ± 0.2 nm, indicating again a high mobility of Ag adatoms on the AZO surface and their consequent coalescence. To summarize, the combination of the synthesis methods (RF sputtering and SCBD) with the substrate role (adatom mobility and surface roughness) is providing a very good playground to modify the deposited grain size and final roughness of the Ag/AZO coatings.

3.2. Wettability Analyses

Contact angles measured for water and diiodomethane and related surface energies of the different coatings deposited on PLA are reported in Table 2 and plotted in Figure 4.

Table 2. Contact angles of water and diiodomethane and surface energy values for coatings deposited on PLA.

Sample	Contact Angle (°)		Surface Energy (mJ/m^2)		
	Water	Diiodomethane	Polar Component	Dispersion Component	Total
AZO	63.6 ± 3.2	55.8 ± 3.0	13.9	31.0	44.9 ± 2.7
Ag(RFS)	80.8 ± 2.3	49.6 ± 2.8	4.3	34.5	38.8 ± 1.8
Ag(SCBD)	91.8 ± 5.5	55.2 ± 2.4	1.6	31.3	32.9 ± 1.9
Ag(RFS)/AZO	93.7 ± 1.9	57.8 ± 3.9	1.4	29.9	31.3 ± 2.3
Ag(SCBD)/AZO	91.3 ± 5.9	55.9 ± 4.1	1.8	30.9	32.7 ± 2.8

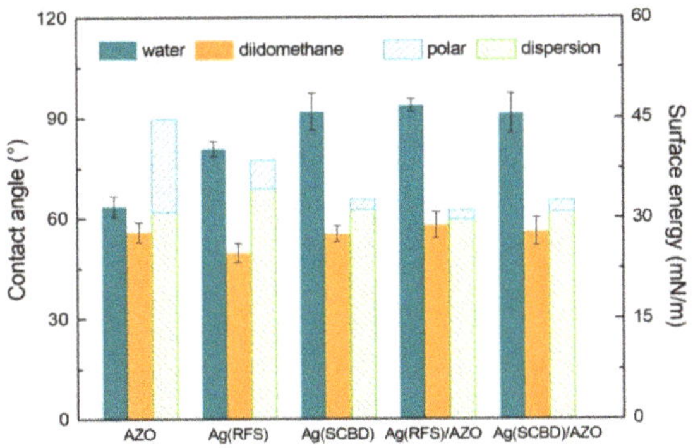

Figure 4. Contact angles (plain bars) and surface energy components (patterned bars) of samples deposited on PLA.

The water contact angle (WCA) values indicate that the water wettability of all the Ag-containing samples is significantly different from that one of the AZO coating, with the latter presenting a WCA of about 64° while the others showing higher WCAs around 80° or slightly above 90°. As a consequence of the Ag addition onto the AZO coating (samples Ag(RFS)/AZO and Ag(SCBD)/AZO), the surface behavior is then turned from moderately hydrophilic to slightly hydrophobic, regardless of the deposition technique used for the Ag functionalization. On the contrary, no particular variation is observed on the wettability with the non-polar solvent diiodomethane, where contact angles values are very similar for all the different coatings. In line with the differences in WCA, the total surface energy of the AZO layer is also different from that one obtained on samples having silver on their surface: the Ag presence on the surface reduces the total surface energy, where this decrease is primarily led by the strong reduction of the polar component (corresponding to the increase of the related contact angle with the polar liquid water).

Since the wettability of a surface is determined by a combination of its physico-chemical characteristics, one of the reasons for the WCA differences between the Ag-containing coatings and the unfunctionalized AZO sample can be related to the different chemical state of the surface (dangling bonds, terminal facets of the particles, oxidation, etc.) due to the presence of the Ag clusters. At the same time, influence of surface morphology can be deduced as well. Indeed, the lower contact angles observed for both water and diiodomethane in sample Ag(RFS) as compared to Ag(SCBD) can be attributed to its lower roughness, since the spread of a liquid droplet on a smooth surface is facilitated with respect to a rougher surface where air can be trapped below the droplet and hinder its spread. On the contrary, the contact angle values obtained on sample Ag(RFS)/AZO increase to values comparable to the analogous sample Ag(SCBD)/AZO, due to the increased roughness when Ag is sputtered on the AZO coating, as described above.

3.3. Antimicrobial Tests

Results of antimicrobial tests performed on the Gram-negative bacterium *E. coli* and Gram-positive bacterium *S. aureus* are shown in Figure 5, reporting the bacterial survival rate at different time intervals for the various coated samples, together with the reference data of the control sample (i.e., uncoated PLA substrate). In order to quantify the net antimicrobial action of the coatings with respect to the control, the antimicrobial activity A of the coated samples was calculated as

$$A = F - G = \log \frac{C_t/C_{t0}}{T_t/T_{t0}} \tag{1}$$

with $F = (\text{Log } C_t - \text{Log } C_{t0})$ and $G = (\text{Log } T_t - \text{Log } T_{t0})$, where C_t and T_t are the CFUs detected at the different considered times t for control substrate and coated samples, respectively, and C_{t0} and T_{t0} are the respective CFUs at the initial time $t = 0$. The antimicrobial activity, whose values are listed in Table 3 and plotted in Figure 6, can be considered good when $A \geq 2$, and excellent when $A \geq 3$.

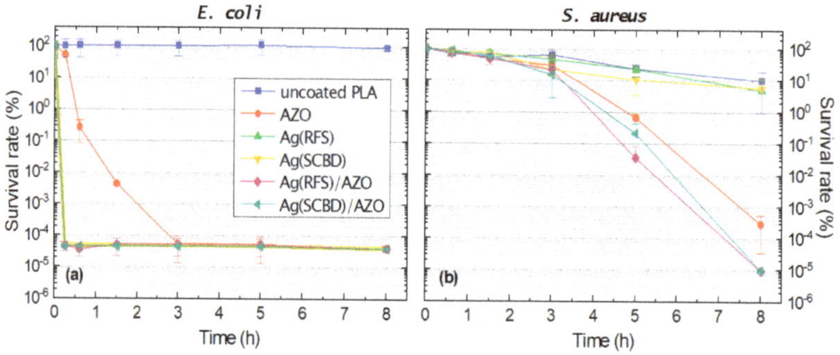

Figure 5. Antibacterial effect of the different samples against (**a**) the Gram-negative bacterium *E. coli* and (**b**) the Gram-positive bacterium *S. aureus*, at different time intervals. The survival rate is intended as the percentage reduction of CFU/mL with respect to $t = 0$. The y axes are in logarithmic scale.

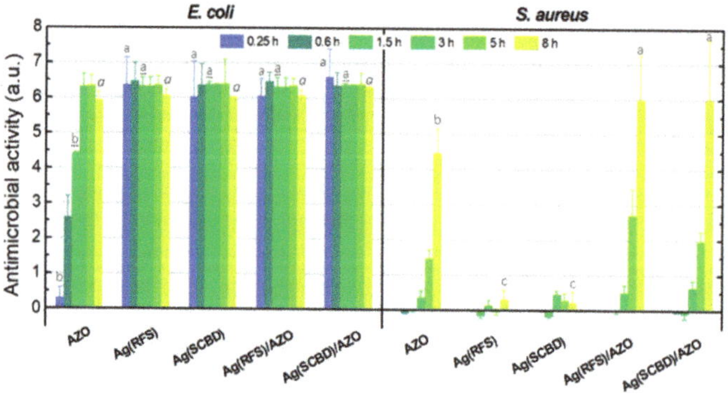

Figure 6. Antimicrobial activity (A) of the different samples against *E. coli* (left) and *S. aureus* (right). Statistical analysis refers at same times and, for each of them, different letters indicate significant differences among coated films ($p < 0.001$). The same character (normal, italic and underscored) is used for each considered time.

Table 3. Antimicrobial activity A against *E. coli* and *S. aureus*, calculated by Equation (1) at different times for samples deposited on PLA. Data are reported as the mean ± standard deviation.

Bacterial Species	Sample	Antimicrobial Activity (A)					
		Treatment Time (h)					
		0.25 *	0.6	1.5	3	5	8
E. coli	AZO	0.30 ± 0.30	2.59 ± 0.60	4.40 ± 0.04	6.30 ± 0.36	6.33 ± 0.28	5.90 ± 0.25
	Ag(RFS)	6.36 ± 0.78	6.47 ± 0.53	6.32 ± 0.26	6.32 ± 0.26	6.40 ± 0.25	6.00 ± 0.18
	Ag(SCBD)	6.03 ± 1.00	6.36 ± 0.60	6.37 ± 0.08	6.37 ± 0.02	6.40 ± 0.70	6.08 ± 0.01
	Ag(RFS)/AZO	6.06 ± 0.50	6.47 ± 0.27	6.32 ± 0.26	6.32 ± 0.26	6.35 ± 0.20	6.05 ± 0.17
	Ag(SCBD)/AZO	6.60 ± 0.80	6.36 ± 0.38	6.37 ± 0.05	6.37 ± 0.02	6.40 ± 0.30	6.30 ± 0.01
S. aureus	AZO	n.d.	−0.08 ± 0.03	0.02 ± 0.08	0.35 ± 0.19	1.45 ± 0.25	4.44 ± 0.69
	Ag(RFS)	n.d.	0.00 ± 0.03	−0.15 ± 0.08	0.13 ± 0.15	−0.05 ± 0.09	0.29 ± 0.26
	Ag(SCBD)	n.d.	0.00 ± 0.02	−0.17 ± 0.04	0.45 ± 0.10	0.28 ± 0.19	0.19 ± 0.33
	Ag(RFS)/AZO	n.d.	0.02 ± 0.02	0.00 ± 0.10	0.49 ± 0.21	2.70 ± 0.73	5.96 ± 1.27
	Ag(SCBD)/AZO	n.d.	−0.04 ± 0.03	−0.10 ± 0.14	0.64 ± 0.20	1.96 ± 0.26	5.98 ± 1.70

* Data at 0.25 h for *S. aureus* were not determined (n.d.) as there was no detectable variation with respect to $t = 0$.

The survival kinetics measured for *E. coli* (Figure 5a) demonstrate a strong bactericidal action of the unfunctionalized AZO coating, with a survival rate reaching the minimum detection limit (~3–5 CFU/mL) after about 3 h, corresponding to a reduction of cellular density higher than 6 Log ($A = 6.3$, i.e., reduction in bacterial population >99.9999 %), that is in line with results reported in our previous work on similar AZO samples [8]. However, the coatings made of only silver on PLA (Ag(RFS) and Ag(SCBD)) clearly evidence even far better behavior against *E. coli*, with an extremely fast action inducing the same reduction of about 6 Log in only 15 min. Then, the composite samples made of Ag-functionalized AZO coatings (Ag(RFS)/AZO and Ag(SCBD)/AZO) follow the same trend of the single Ag coatings (the survival rate curves of the Ag/AZO coatings and those of the Ag coatings are overlapped in Figure 5), indicating that, in the examined time range (minimum of 15 min), their antibacterial response is driven by the activity of silver.

In contrast, an opposite behavior is observed in the kinetics recorded for the Gram-positive bacterium *S. aureus* (Figure 5b), where the simple silver coatings demonstrate a very low antibacterial efficacy, while the response of the composite AZO + Ag coatings is mainly driven by the AZO contribution. As a preliminary observation, a weak antimicrobial influence on *S. aureus* can be detected at long time for the bare PLA substrate (uncoated control sample) too, inducing about 1 Log reduction of the bacterial colony counts after 8 h. This minimal effect can be probably ascribed to a rather low biocide action caused by the coarseness of the PLA surface, inducing mechanical detrimental effects (e.g., cell deformations and wall abrasion), and to environmental stress induced by the oligotrophic conditions of the incubation medium used during the microbiological tests. For longer exposition times either or both the factors could then result in a slight bacterial reduction. Such behavior is not observed on *E. coli* in the examined time range because, as a microbiological indicator of water quality, it is more stable in environments poor in organic substances compared to a pathogenic bacterium such as *S. aureus* [25,26].

Analyzing the results of the coated samples, the antibacterial action against *S. aureus* is shown to be slower than *E. coli* for all the films, as clearly evidenced in Figures 5 and 6. In particular, the activity A of the single AZO coating on *S. aureus* is moderate ($A < 2$) up to about 5 h, increasing to a value around 4.4 at 8 h, whereas on *E. coli* the same activity level of $A = 4.4$ is obtained after only 1.5 h and the coating exhibits a complete bactericidal action ($A > 6$) at 3 h. Generally, Gram-positive species like *S. aureus* are considered to have higher susceptibility to ZnO than Gram-negative ones like *E. coli*, due to the differences in thickness, complexity, and composition of cell membranes in the two families, as well as difference in their intracellular content, thus inducing different sensitivity to active species such as metal ions and reactive oxygen species [27]. However, it should be noted that, differently from *E. coli*, *S. aureus* tends to form multicellular aggregates that take longer to be destroyed, therefore high bacterial suppression can be reached after longer exposure times with the active material, as shown in our previous work [8], thus explaining the slower biocidal effect observed on *S. aureus* at short periods.

Contrarily to what observed on *E. coli*, the single Ag coatings on PLA (Ag(RFS) and Ag(SCBD)) demonstrate a very low antimicrobial effect on *S. aureus* even at long times, with a response that is comparable to the bare PLA substrate, resulting in activity values A as low as ~0.2–0.3 at 8 h. The limited efficacy of Ag against *S. aureus* in comparison to *E. coli* observed in our experiments is in line with previous findings [11,28,29], and it was related to the different electrostatic interaction with the charged cell walls of the two bacterial species.

Concerning the composite coatings made of Ag-functionalized AZO layers, the trend of survival rate for *S. aureus* is found to be similar to that one of the single AZO coating, with a moderate improvement introduced by the silver presence. To our knowledge, there is no other report in literature about the antimicrobial properties of Ag-functionalized AZO coatings, thus we can only draw a general comparison with analogous materials made of combination of Ag with ZnO. The antibacterial behavior in our samples is similar or appears to be even faster, especially for *E. coli*, when compared to other kinds of Ag-modified ZnO nanomaterials (see e.g., [30–32] and references therein), observing the different activity of the two individual materials on Gram-negative and Gram-positive species and the beneficial effects of joining them together. In particular, in [31], Ag was shown to be less effective on *S. aureus* than *E. coli*, and the inhibitory effect was enhanced in the Ag/ZnO heterostructure NPs against both species, presenting a better response against the latter, which is in line with the behavior demonstrated by our samples too. Anyhow, it must be underlined that these properties are strongly dependent on a series of experimental factors, such as materials composition and concentration, form (different kinds of nanostructures), crystal structure, substrates, synthesis procedures, methodologies and parameters used for the antimicrobial tests, etc., thus making it impossible to definitely derive a reliable comparison. Our Ag/AZO coatings are able to reach a complete suppression of *E. coli* in only 15 min, while almost complete bactericide action against *S. aureus* is induced in about 8 h (around 6 Log of CFU reduction). This suppression rate for *S. aureus* in the hybrid coatings corresponds to an activity value that is about 1.3 times higher than the single AZO coating at the same exposure time. Considering the minor antibacterial action revealed for the individual Ag coatings against *S. aureus* as described above, the improvement observed for this bacterial species in the composite coatings can be only partially ascribed to the secondary additional contribution of silver on AZO. As a further enhancing factor, the increased surface roughness in the AZO + Ag coatings with respect to the single AZO film might also play a significant role in increasing the antimicrobial efficacy. Indeed, higher surface roughness has been reported as a key factor to increase the bactericidal action [8,33], in consequence of the enlarged effective contact area between the rough surface and the surrounding environment, promoting microbial adhesion through electrostatic interactions with the surface, and resultant enhanced bactericidal effects such as structural deformation of the microorganisms, abrasive damage of cell membranes, and fostered transfer of active species (metal ions and ROS) from the coating surface to the bacterial cells.

For both bacterial species here considered, no significant difference in the antimicrobial characteristics can be evidenced when comparing analogous samples where silver has been deposited by the two different techniques (RFS or SCBD). The bacterial suppression rate and related antimicrobial activities appear the same regardless of the process used for the Ag deposition in the two sets of samples, pointing out no substantial influence of the used deposition technique on the antimicrobial performance under the experimental conditions and time range here used. More complex experimental analyses should be needed in order to discern for possible fine differences in the antimicrobial response of these samples, especially in the case of *E. coli* where the drop of CFU counts is extremely rapid.

As an overall consideration, the Ag-functionalized AZO coatings on PLA here developed are shown to possess an excellent antibacterial activity against both considered microbial species, with an extremely rapid complete suppression of *E. coli* in a few minutes and a complete suppression of *S. aureus* in a few hours. Thanks to the preferential action of each material against one species or another (predominant efficacy of Ag on *E. coli* and of AZO on *S. aureus* in our samples) it is then possible to obtain a composite coating exploiting the combined performance of both materials together, therefore being capable to simultaneously and efficiently strike both kinds of bacteria.

4. Conclusions

Ecofriendly antibacterial films made of Ag-functionalized Al-doped ZnO nanostructured coatings deposited on bioplastic PLA films were developed with the aim to find new solutions to contrast the diffusion of infections and diseases. Ag-functionalization was evaluated by two different deposition methods—RF sputtering and supersonic cluster beam deposition—as a way to tune the surface features (roughness and particle size) of the final Ag/AZO composite materials. Surface wettability was turned from moderately hydrophilic to slightly hydrophobic by the addition of Ag onto the AZO layer, together with some influence of surface roughness. The antimicrobial action of the Ag-functionalized AZO coatings to the Gram-negative *E. coli* was shown to be led by the Ag presence, while the response against the Gram-positive *S. aureus* was mainly driven by the AZO material. The combination of these materials together enhanced the biocide action, leading to an extremely fast total suppression of *E. coli* within few minutes, while *S. aureus* required some hours, due to the dissimilar bacterial walls and consequent interactions of the two species with the active materials, as well as to different tendency to form multicellular aggregates.

Thanks to the preferential bactericide action of Ag and AZO on the different bacterial species, the hybrid Ag/AZO coatings here developed are then able to exploit a synergistic action, providing simultaneous superior antimicrobial performance against both species with respect to the individual material deposits. They can therefore represent a promising solution to provide broader spectrum of action against different kinds of bacteria for different possible applications, like in food packaging and biomedical sectors. In perspective of these final applications, further insights in the evaluation of the active species releasing properties of these composite coatings can be required to get a deeper understanding of the mechanisms of action in damaging the cytoplasmic membrane of the bacterial cells and a correct evaluation of possible toxicity concerns for human cells.

Author Contributions: Conceptualization, D.V. and L.G.; Methodology, D.V., L.T., G.V., E.P., F.B., E.C., and L.G.; Investigation, all authors; Writing—original draft preparation, D.V. and L.G.; Writing—review and editing, all authors. All authors have read and agreed to the published version of the manuscript.

Funding: This work was supported by MIUR-FARB 2018–2019 funding from Università di Salerno (G.V., E.P.). L.G., E.C., L.C. acknowledge that this research was partially funded by Università Cattolica del Sacro Cuore through D.2.2 and D.3.1 grants.

Conflicts of Interest: The authors declare no conflict of interest.

References

1. Wille, I.; Mayr, A.; Kreidl, P.; Bruhwasser, C.; Hinterberger, G.; Fritz, A.; Posch, W.; Fuchs, S.; Obwegeser, A.; Orth-Holler, D.; et al. Cross-sectional point prevalence survey to study the environmental contamination of nosocomial pathogens in intensive care units under real-life conditions. *J. Hosp. Infect.* **2018**, *98*, 90–95. [CrossRef] [PubMed]
2. Wei, T.; Tang, Z.; Yu, Q.; Chen, H. Smart antibacterial surfaces with switchable bacteria-killing and bacteria-releasing capabilities. *ACS Appl. Mater. Interfaces* **2017**, *9*, 37511–37523. [CrossRef] [PubMed]
3. Wang, L.; Hu, C.; Shao, L. The antimicrobial activity of nanoparticles: Present situation and prospects for the future. *Int. J. Nanomed.* **2017**, *12*, 1227–1249. [CrossRef] [PubMed]
4. Kim, I.; Viswanathan, K.; Kasi, G.; Thanakkasaranee, S.; Sadeghi, K.; Seo, J. ZnO nanostructures in active antibacterial food packaging: Preparation methods, antimicrobial mechanisms, safety issues, future prospects, and challenges. *Food Rev. Int.* **2020**. [CrossRef]
5. Makvandi, P.; Wang, C.; Zare, E.; Borzacchiello, A.; Niu, L.; Tay, F. Metal-based nanomaterials in biomedical applications: Antimicrobial activity and cytotoxicity aspects. *Adv. Funct. Mater.* **2020**, *30*, 1910021. [CrossRef]

6. Valerini, D.; Tammaro, L.; Di Benedetto, F.; Vigliotta, G.; Capodieci, L.; Terzi, R.; Rizzo, A. Aluminum-doped zinc oxide coatings on polylactic acid films for antimicrobial food packaging. *Thin Solid Film.* **2018**, *645*, 187–192. [CrossRef]
7. Saxena, V.; Chandra, P.; Pandey, L. Design and characterization of novel Al-doped ZnO nanoassembly as an effective nanoantibiotic. *Appl. Nanosci.* **2018**, *8*, 1925–1941. [CrossRef]
8. Valerini, D.; Tammaro, L.; Villani, F.; Rizzo, A.; Caputo, I.; Paolella, G.; Vigliotta, G. Antibacterial Al-doped ZnO coatings on PLA films. *J. Mater. Sci.* **2020**, *55*, 4830–4847. [CrossRef]
9. Huang, J.; Li, X.; Zhou, W. Safety assessment of nanocomposite for food packaging application. *Trends Food Sci. Technol.* **2015**, *45*, 187–199. [CrossRef]
10. Wegner, K.; Piseri, P.; Tafreshi, H.; Milani, P. Cluster beam deposition: A tool for nanoscale science and technology. *J. Phys. D Appl. Phys.* **2006**, *39*, R439–R459. [CrossRef]
11. Cavaliere, E.; De Cesari, S.; Landini, G.; Riccobono, E.; Pallecchi, L.; Rossolini, G.; Gavioli, L. Highly bactericidal Ag nanoparticle films obtained by cluster beam deposition. *Nanomed.-Nanotechnol. Biol. Med.* **2015**, *11*, 1417–1423. [CrossRef] [PubMed]
12. Cavaliere, E.; Benetti, G.; Van Bael, M.; Winckelmans, N.; Bals, S.; Gavioli, L. Exploring the optical and morphological properties of Ag and Ag/TiO_2 nanocomposites grown by supersonic cluster beam deposition. *Nanomaterials* **2017**, *7*, 442. [CrossRef] [PubMed]
13. Benetti, G.; Cavaliere, E.; Banfi, F.; Gavioli, L. Antimicrobial nanostructured coatings: A gas phase deposition and magnetron sputtering perspective. *Materials* **2020**, *13*, 784. [CrossRef] [PubMed]
14. Owens, D.K.; Wendt, R.C. Estimation of the surface free energy of polymers. *J. Appl. Polym. Sci.* **1969**, *13*, 1741–1747. [CrossRef]
15. Ström, G.; Fredriksson, M.; Stenius, P. Contact angles, work of adhesion, and interfacial tensions at a dissolving hydrocarbon surface. *J. Colloid Interface Sci.* **1987**, *119*, 352–361. [CrossRef]
16. Pantani, R.; Gorrasi, G.; Vigliotta, G.; Murariu, M.; Dubois, P. PLA-ZnO nanocomposite films: Water vapor barrier properties and specific end-use characteristics. *Eur. Polym. J.* **2013**, *49*, 3471–3482. [CrossRef]
17. Milani, P.; Iannotta, S. *Cluster Beam Synthesis of Nanostructured Materials*, 1st ed.; Springer: Berlin/Heidelberg, Germany, 1999.
18. Barborini, E.; Kholmanov, I.; Piseri, P.; Ducati, C.; Bottani, C.; Milani, P. Engineering the nanocrystalline structure of TiO_2 films by aerodynamically filtered cluster deposition. *Appl. Phys. Lett.* **2002**, *81*, 3052–3054. [CrossRef]
19. Benetti, G.; Cavaliere, E.; Brescia, R.; Salassi, S.; Ferrando, R.; Vantomme, A.; Pallecchi, L.; Pollini, S.; Boncompagni, S.; Fortuni, B.; et al. Tailored Ag-Cu-Mg multielemental nanoparticles for wide-spectrum antibacterial coating. *Nanoscale* **2019**, *11*, 1626–1635. [CrossRef]
20. Benetti, G.; Cavaliere, E.; Canteri, A.; Landini, G.; Rossolini, G.; Pallecchi, L.; Chiodi, M.; Van Bael, M.; Winckelmans, N.; Bals, S.; et al. Direct synthesis of antimicrobial coatings based on tailored bi-elemental nanoparticles. *APL Mater.* **2017**, *5*, 036105. [CrossRef]
21. Benetti, G.; Caddeo, C.; Melis, C.; Ferrini, G.; Giannetti, C.; Winckelmans, N.; Bals, S.; Van Bael, M.; Cavaliere, E.; Gavioli, L.; et al. Bottom-up mechanical nanometrology of granular ag nanoparticles thin films. *J. Phys. Chem. C* **2017**, *121*, 22434–22441. [CrossRef]
22. Benetti, G.; Gandolfi, M.; Van Bael, M.; Gavioli, L.; Giannetti, C.; Caddeo, C.; Banfi, F. Photoacoustic sensing of trapped fluids in nanoporous thin films: Device engineering and sensing scheme. *ACS Appl. Mater. Interfaces* **2018**, *10*, 27947–27954. [CrossRef] [PubMed]
23. Depla, D.; Mahieu, S.; Greene, J.E. Chapter 5—Sputter Deposition Processes. In *Handbook of Deposition Technologies for Films and Coatings*, 3rd ed.; Martin, P.M., Ed.; William Andrew Publishing: Boston, MA, USA, 2010; pp. 253–296.
24. Maréchal, N.; Quesnel, E.; Pauleau, Y. Silver thin films deposited by magnetron sputtering. *Thin Solid Film.* **1994**, *241*, 34–38. [CrossRef]
25. Flint, K.P. The long-term survival of *Escherichia coli* in river water. *J. Appl. Bacteriol.* **1987**, *63*, 261–270. [CrossRef] [PubMed]
26. Watson, S.; Clements, M.; Foster, S. Characterization of the starvation-survival response of *Staphylococcus aureus*. *J. Bacteriol.* **1998**, *180*, 1750–1758. [CrossRef]
27. Gold, K.; Slay, B.; Knackstedt, M.; Gaharwar, A. Antimicrobial activity of metal and metal-oxide based nanoparticles. *Adv. Ther.* **2018**, *1*, 1700033. [CrossRef]

28. Kim, J.; Kuk, E.; Yu, K.; Kim, J.; Park, S.; Lee, H.; Kim, S.; Park, Y.; Park, Y.; Hwang, C.; et al. Antimicrobial effects of silver nanoparticles. *Nanomed.-Nanotechnol. Biol. Med.* **2007**, *3*, 95–101. [CrossRef]
29. Mendez-Pfeiffer, P.; Urzua, L.; Sanchez-Mora, E.; Gonzalez, A.; Romo-Herrera, J.; Arciniega, J.; Morales, L. Damage on *Escherichia coli* and *Staphylococcus aureus* using white light photoactivation of Au and Ag nanoparticles. *J. Appl. Phys.* **2019**, *125*, 213102. [CrossRef]
30. Lu, W.; Liu, G.; Gao, S.; Xing, S.; Wang, J. Tyrosine-assisted preparation of Ag/ZnO nanocomposites with enhanced photocatalytic performance and synergistic antibacterial activities. *Nanotechnology* **2008**, *19*, 445711. [CrossRef]
31. Zhang, Y.; Gao, X.; Zhi, L.; Liu, X.; Jiang, W.; Sun, Y.; Yang, J. The synergetic antibacterial activity of Ag islands on ZnO (Ag/ZnO) heterostructure nanoparticles and its mode of action. *J. Inorg. Biochem.* **2014**, *130*, 74–83. [CrossRef]
32. Agnihotri, S.; Bajaj, G.; Mukherji, S. Arginine-assisted immobilization of silver nanoparticles on ZnO nanorods: An enhanced and reusable antibacterial substrate without human cell cytotoxicity. *Nanoscale* **2015**, *7*, 7415–7429. [CrossRef]
33. Cao, F.; Zhang, L.; Wang, H.; You, Y.; Wang, Y.; Gao, N.; Ren, J.; Qu, X. Defect-rich adhesive nanozymes as efficient antibiotics for enhanced bacterial inhibition. *Angew. Chem.-Int. Ed.* **2019**, *58*, 16236–16242. [CrossRef] [PubMed]

Publisher's Note: MDPI stays neutral with regard to jurisdictional claims in published maps and institutional affiliations.

coatings

Article

Fabrication of Zinc Oxide-Xanthan Gum Nanocomposite via Green Route: Attenuation of Quorum Sensing Regulated Virulence Functions and Mitigation of Biofilm in Gram-Negative Bacterial Pathogens

Fohad Mabood Husain [1,*], Imran Hasan [2], Faizan Abul Qais [3], Rais Ahmad Khan [4], Pravej Alam [5] and Ali Alsalme [4,*]

1 Department of Food Science and Nutrition, College of Food and Agriculture Sciences, King Saud University, Riyadh 11451, Saudi Arabia
2 Department of Chemistry, Chandigarh University, Mohali 140413, India; imranhasan98@gmail.com
3 Department of Agricultural Microbiology, Faculty of Agricultural Sciences, Aligarh Muslim University, Aligarh 202002, India; faizanabulqais@gmail.com
4 Department of Chemistry, King Saud University, Riyadh 11451, Saudi Arabia; krais@ksu.edu.sa
5 Department of Biology, College of Science and Humanities in Al-Kharj, Prince Sattam bin Abdulaziz University, Al-Kharj 11942, Saudi Arabia; alamprez@gmail.com
* Correspondence: fhusain@ksu.edu.sa (F.M.H.); aalsalme@ksu.edu.sa (A.A.)

Received: 14 November 2020; Accepted: 3 December 2020; Published: 5 December 2020

Abstract: The unabated abuse of antibiotics has created a selection pressure that has resulted in the development of antimicrobial resistance (AMR) among pathogenic bacteria. AMR has become a global health concern in recent times and is responsible for a high number of mortalities occurring across the globe. Owing to the slow development of antibiotics, new chemotherapeutic antimicrobials with a novel mode of action is required urgently. Therefore, in the current investigation, we green synthesized a nanocomposite comprising zinc oxide nanoparticles functionalized with extracellular polysaccharide xanthan gum (ZnO@XG). Synthesized nanomaterial was characterized by structurally and morphologically using UV-visible spectroscopy, XRD, FTIR, BET, SEM and TEM. Subinhibitory concentrations of ZnO@XG were used to determine quorum sensing inhibitory activity against Gram-negative pathogens, *Chromobacterium violaceum*, and *Serratia marcescens*. ZnO@XG reduced quorum sensing (QS) regulated virulence factors such as violacein (61%), chitinase (70%) in *C. violaceum* and prodigiosin (71%) and protease (72%) in *S. marcescens* at 128 µg/mL concentration. Significant ($p \leq 0.05$) inhibition of biofilm formation as well as preformed mature biofilms was also recorded along with the impaired production of EPS, swarming motility and cell surface hydrophobicity in both the test pathogens. The findings of this study clearly highlight the potency of ZnO@XG against the QS controlled virulence factors of drug-resistant pathogens that may be developed as effective inhibitors of QS and biofilms to mitigate the threat of multidrug resistance (MDR). ZnO@XG may be used alone or in combination with antimicrobial drugs against MDR bacterial pathogens. Further, it can be utilized in the food industry to counter the menace of contamination and spoilage caused by the formation of biofilms.

Keywords: xanthan gum; zinc oxide; nanocomposite; quorum sensing; biofilm; virulence; *S. marcescens*; *C. violaceum*

1. Introduction

Xanthan gum is a natural anionic extracellular polysaccharide. The non-toxic and biocompatible nature of this polymer makes it quite useful for the food sector [1]. The inorganic particles, such as metals, can easily be adsorbed onto it to form a stable emulsion without altering the interfacial tension. This is a Food and Drug Administration (FDA)-approved biopolymer for the food industry [2,3]. Zinc is an essential micronutrient and therefore extensively prescribed for its nutritional supplement in case of deficiency for human health. As zinc is biocompatible, there is not much risk associated with public health and is used as food coating materials [4]. The unique property of zinc oxide nanoparticles is exploited as promising antibacterial, antibiofilm, or antivirulence candidates (Al-Shabib et al., 2018; Sirelkhatim et al., 2015, [5,6]).

Tremendous growth and spread of multidrug resistance (MDR) among microbial pathogens have become a global concern for human health countries [7]. The worldwide deaths of human-caused by antimicrobial resistance (AMR) is a major contributor to global mortality after cancer and cardiovascular diseases [8]. The problem caused by AMR has reached an alarming situation, and if not action is taken, it is expected to cause more global mortality than cancer by 2050 [9,10]. The infections caused by MDR pathogens are an epidemiological concern and worsen the treatment of infectious diseases by diminishing the therapeutic effectiveness of antibiotics [11,12]. Moreover, AMR is not only problematic for public health but also poses an extra burden on the environment and livestock. The first half 20th century is regarded as the golden era for the discovery and development of antibiotics. Nearly 70% of all antibiotics used so far were discovered by 1960. After this, there was poor progress in the antibiotic drug discovery and in the last four decades, only a few antibiotics exhibiting a novel mechanism of action are discovered [13]. The injudicious use of antibiotics in public health and health and environment creates a selection pressure that results in the development of AMR among microbial pathogens [14,15].

The risk of AMR development is so much that even the antibiotics of the last generation cannot be entrusted for prolonged applications. Hence, there are two major issues associated with the discovery or development of antibacterials; the first is to find or make new chemotherapeutic antimicrobials with novel modes of action, and the second is to minimize the risk of development of AMR against the discovered antimicrobials. This has led the researchers to focus on the development of alternative anti-infective strategies to combat AMR.

Quorum sensing (QS) is a communication system occurring via chemical signaling that operates as a function of the density of the bacterial population [16]. These chemical signal molecules are called autoinducers (AIs). Certain phenotypes of bacteria are only expressed when their population has reached a certain threshold value. Many clinically important traits of bacteria, such as virulence production, biofilms formation, expression of drug-resistant genes, are controlled via QS [17]. As the bacterial population increases, the concentration of AIs increases and triggers the expression of QS-regulated genes by transcriptional regulation when it reaches a certain limit [18]. Biofilms are the microbes residing in a biopolymeric matrix. Earlier, it was thought that bacteria live in a planktonic state. However, it was discovered that most of the bacteria form complex structures called biofilms [19]. A large number of bacterial infections are encouraged or caused by the formation of biofilms by the pathogenic or opportunistic pathogens [20,21]. One of the novel strategies in antibacterial drug discovery is to selectively target the bacterial quorum sensing and biofilms. Among new drug candidates, natural/biocompatible products and nanoparticles have proven as an important therapeutic antimicrobial alternative (Husain et al., 2019; Qais et al., 2018, 2019, [22–24]).

In this study, green synthesis of zinc oxide xanthan gum (ZnO@XG) nanocomposite was done. The antiquorum-sensing potential of ZnO@XG was tested against biosensor strains of Gram-negative bacteria. The QS controlled violacein production, and chitinase activity was tested in *C. violaceum*. QS regulated virulence traits of *S. marcescens* such as prodigiosin production, and protease production were also assessed at subinhibitory concentrations of ZnO@XG. Further, the effect on biofilm formation, preformed biofilm and factors such as EPS production, swarming motility, and cell surface hydrophobicity that contribute to the development of biofilm against both the bacteria was also studied. This is probably the first study assessing the quorum sensing and biofilm inhibitory potential of green synthesized polysaccharide-zinc oxide nanocomposite.

2. Materials and Experimental Methods

2.1. Chemicals

Xanthan gum extracted from *Xanthomonas Campestris* biological grade was purchased from Sigma-Aldrich (Bangalore, India). Zinc nitrate (Zn $(NO_3)_2{\cdot}6H_2O$ white crystals) and sodium hydroxide (NaOH) were purchased from Merck (Mumbai, India). All the chemical materials were used without any purification or refinement. All the aqueous solutions were prepared using deionized water.

2.2. One-Pot Green Synthesis of ZnO@XG Nanoparticles

The Nanoparticles were consolidated by using a chemical coprecipitation scheme using an ecological green route [25]. In a 3 necked round-bottomed flask, a 100 mL solution of 0.45 M Zn $(NO_3)_2{\cdot}6H_2O$ was taken and placed under magnetic stirring (900 rpm) for 30 min to obtain homogeneity. A solution of 2.3% xanthan gum was prepared by dissolving 2.3 g powder in 100 mL deionized water with vigorous stirring at 40 °C for 2 h to obtain a complete bubble-free homogeneous solution. Now the blended solution of xanthan gum with 20 mL of 0.1 M NaOH solution was added drop-by-drop to the aqueous ionic solution of Zn^{2+} in order to extend the reducing character of xanthan gum to the bulk of Zn^{2+}. The mixture remained on vigorous stirring under observation at 40 °C, and the progress of the reaction was checked by taking small aliquots of the reaction mixture at different time intervals to verify using UV-vis spectroscopy (Figure 1). Finally, after 8 h, a white precipitated colloid was obtained from which the product was isolated using a centrifuge (REMI rpm 8500). The product was squeezed using deionized water seven to eight times for the efficient removal of nonreactive species and dried in a hot air oven for 3 h at 60 °C.

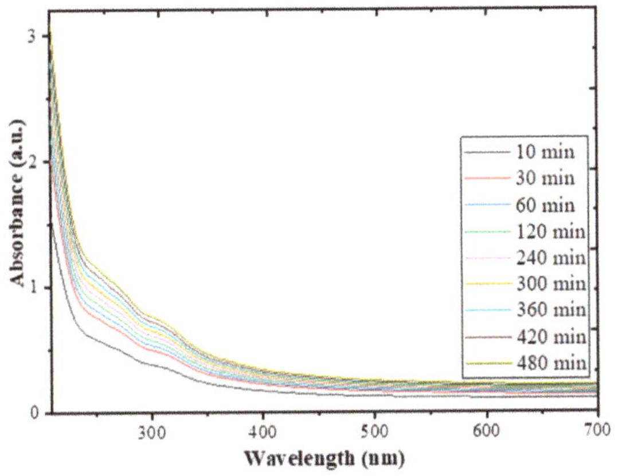

Figure 1. Time-dependent UV-vis spectra for green synthesized ZnO@XG nanocomposite.

2.3. Analytical Techniques Used for Characterization

The prepared nanocomposite and its crystal structure were characterized by several characterization techniques such as FTIR, XRD, SEM-EDX, TEM, BET and UV-Vis. The type of bonding and functional groups present in the synthesized material was investigated by using Fourier-transform infrared spectroscopy (FTIR) PerkinElmer (PE1600, PerkinElmer, Waltham, MA, USA) in the frequency range of 400–4000 cm^{-1} with transmission mode. The crystal phases of the synthesized material were collected on an X-ray diffractometer (A Rigaku Ultima 1 V, Woodlands, TX, USA). The morphologies of the sample were analyzed by using scanning electron microscopy (SEM; JEOL GSM 6510LV, JEOL, Tokyo, Japan). The elemental size and dispensation of the sample were examined by JEM 2100 (Tokyo, Japan) transmission electron microscopy (TEM). For the analysis of aliquots of ZnO@XG samples during the synthesis process, Shimadzu UV-1900 UV-vis double beam spectrophotometer was taken into consideration. The specific surface areas of the synthesized material were tested on Micromeritics Tristar II (Micromeritics, Atlanta, GA, USA) and calculated using the Brunauer–Emmett–Teller (BET) method.

2.4. Bacterial Strains

Chromobacterium violaceum ATCC 12472 and *Serratia marcescens* ATCC 13880 were used to evaluate the QS and biofilm inhibitory property of the synthesized nanocomposites. Stock cultures of all the test bacteria were maintained on nutrient-agar under refrigeration and subcultured in Luria–Bertani (LB) broth for 24 h. Overnight culture (1%) was added to the fresh LB medium [the final optical density (OD) was adjusted to 0.1 at 600 nm].

2.5. Determination of Minimum Inhibitory Concentration (MIC)

The minimum inhibitory concertation of ZnO@XG against test bacteria was assessed by microbroth dilution assay as described previously [26,27]. Briefly, bacteria were cultured in the presence of different concentrations (1024–0.125 μg/mL) of ZnO@XG. Post incubation, TTC (10 μL) was added to each well and incubated at room temperature for 20 min to observe a change in color. The lowest concentration at which the development of pink color was not observed was termed as the MIC.

2.6. Violacein Inhibition Assay

Violacein pigment production was quantified spectrophotometrically using the previously described protocol [28]. Briefly, *C. violaceum* 12,472 (CV12472) was grown overnight in the absence and presence of sub-MICs (16–128 μg/mL) of ZnO@XG at 30 °C. Post incubation, 1 mL culture was centrifuged (10,000 rpm) for 5 min, and violacein pigment was extracted from the pellet using 1 mL DMSO. The extracted mixture was centrifuged to pellet out the bacterial cells. The optical density (OD) of supernatant was recorded at 585 nm using a UV-2600 spectrophotometer (Shimadzu, Kyoto, Japan).

2.7. Chitinolytic Activity

A dye-release assay involving chitin azure was employed to quantify the chitinase produced by CV12472 [29,30]. Briefly, 100 mL bacterial supernatant obtained from treated and untreated bacteria was mixed with 1 mL phosphate buffer containing 10 mg chitin azure and incubated overnight at 37 °C. The insoluble substrate was pelleted out, and absorbance was read at 585 nm.

2.8. Prodigiosin Assay

Production of red-colored prodigiosin in ZnO@XG treated and untreated *S. marcescens* was quantified using the method described previously [31]. Briefly, pellets obtained from the overnight grown cultures of *S. marcescens* were resuspended in 1 mL acidified ethanol and vortexed vigorously. The mixture was centrifuged at 13,000 rpm for 5 min, and the resulting supernatant was read for absorbance at 534 nm.

2.9. Protease Assay

The proteolytic activity in cell-free supernatant of *S. marcescens* was assessed using azocasein as the substrate [32]. ZnO@XG treated and untreated cultures of *S. marcescens* were grown overnight on shaking, and cell-free supernatants (CFS) were collected by centrifugation (7000× *g* for 10 min). Subsequently, CFS (75 μL) was added to 2% azocasein (125 μL) in 0.25 mol Tris (pH 8.0) and incubated for a half-hour at 37 °C. The reaction was stopped with the addition of 10% Trichloroacetic acid, and the mixture was centrifuged for 10 min. The absorbance of the resultant supernatant was measured at 440 nm.

2.10. Biofilm Inhibition Studies

2.10.1. Microtiter Plate Assay

Overnight grown test pathogens *C. violaceum* and *S. marcescens* were diluted in wells containing fresh Tryptic soy broth, sub-MICs of ZnO@XG was added to wells and incubated at 37 °C for 24 h duration. Wells were decanted to get rid of unattached cells and washed with sterile water. Then bound cells were stained with crystal violet and incubated. After 15 min incubation, the dye was removed from wells, and thorough washing was done to remove excess stain. The absorbance of each well was measured at 585 nm to quantify the biofilm inhibition (Al-Shabib et al., 2020, [33]).

2.10.2. Confocal Laser Scanning Microscopic (CLSM) Visualization of Biofilm Structure

CLSM analysis of biofilm of ZnO@XG treated and untreated test pathogens formed on glass coverslips were performed using the protocol described previously (Al-Shabib et al., 2020, [34]). Biofilm was developed on glass coverslips placed in 24 well tissue culture plates as described above. CLSM imaging was done on coverslips stained with 0.1% acridine orange in the dark under JEOL-JSM 6510 LV confocal laser scanning microscope.

2.10.3. Quantification of EPS

EPS was quantified from test pathogens grown in the presence and absence of sub-MICs of ZnO@XG at 30 °C for 24 h. Incubated cultures were centrifuged, ice-cold ethanol (3 volumes) was added to the resultant supernatant, and the mixture was left at 4 °C for 18 h. Subsequently, the mixture was centrifuged, and the pellet was dissolved in 1 mL deionized water [35]. EPS was quantified using the standard method for the estimation of sugars, as described previously [36].

2.10.4. Swarming Motility

Effect of sub-MICs of ZnO@XG on the swarming motility of *C. violaceum* and *S. marcescens* was determined by point inoculating the bacteria on LB soft agar plates and incubating the plates at 30 °C for 18 h. Briefly, LB soft agar plates (% agar *w*/*v*) were prepared, containing ZnO@XG. No treatment was given to control plates. Bactria were point-inoculated in soft agar plates and incubated under static conditions. Post incubation, the diameter of the swarm was measured in treated and control plates to evaluate the inhibition of swarming motility [37].

2.10.5. Microbial Adhesion to Hydrocarbon (MATH) Assay

MATH assay was used to evaluate the effect of ZnO@XG on cell surface hydrophobicity of the test pathogens [38]. Optical density (OD) of treated and untreated bacteria was recorded at 600 nm. Then, toluene was added to each set and vortexed for 10 min, and the mixture was left for separation of phases. The aqueous phase was collected, and the absorbance was recorded at 600 nm. Percent of hydrophobicity was calculated using the following formulae:

$$\%\ \text{hydrophobicity} = [1 - \text{OD after vortexing}/\text{OD before vortexing}] \times 100\%$$

2.11. Disruption of Preformed Biofilms

Biofilms were allowed to develop for 24 h in the wells of microtiter plates. Non-adhering cells were washed away, new sterile TSB with or without sub-MICs of ZnO@XG was supplemented to each well and MTP was incubated at 37 °C for 24 h. Then, unbound cells were removed by washing and adhering cells were stained with crystal violet. Absorbance was recorded at 585 nm, as described earlier (Al-Shabib et al., 2020, [39]).

3. Results and Discussion

Nanocomposites of zinc oxide and xanthan gum were synthesized according to the scheme depicted in Figure 2. Synthesized nanomaterial was characterized by structurally and morphologically using various spectroscopic and microscopic techniques

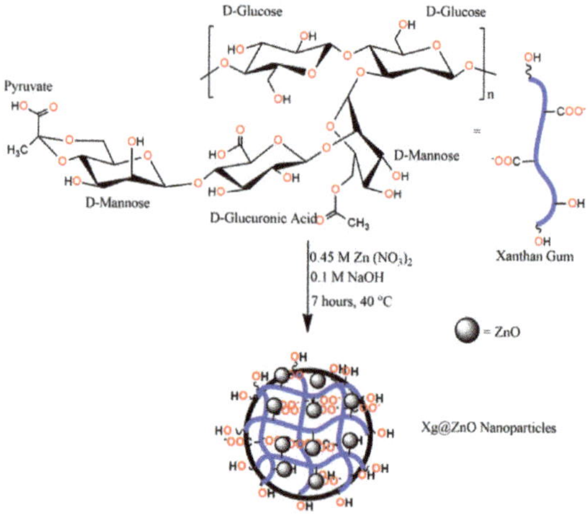

Figure 2. Proposed scheme for the synthesis of ZnO@XG nanocomposite.

3.1. FTIR

The Fourier-transform infrared spectra of xanthan gum and ZnO@XG NPs are displayed in Figure 3. The FTIR spectra of XG (Figure 3a) are demonstrated the peak at 3417 (–OH stretching), 2932 (aliphatic –CH stretching), 1736 (–C=O stretching), 1614, 1414 cm^{-1} (COO– symmetric and asymmetric stretching) and the peaks between 1049–1249 cm^{-1} (pyranoid C-O-C ring stretching) [40]. The FTIR spectra of ZnO@XG NPs in Figure 3b represents all the characteristic peaks from XG and ZnO with vibrational frequency, e.g., 606, 774 (Zn–O bond stretching), 1049–1221 (C–O–C XG pyranoid ring), 1409, 1596 (COO– symmetric and asymmetric stretching), 1736 (–C=O stretching), 2923, 2850 (C–H aliphatic stretching of XG), 3307 cm^{-1} (–OH stretching). The shifting in the carboxylic acid vibrational frequency suggests that the reduction as well stabilization of the Zn^{2+} into ZnO was done through the donation of electrons (lone pairs) from oxygen from an XG–O–Zn type lattice [41,42].

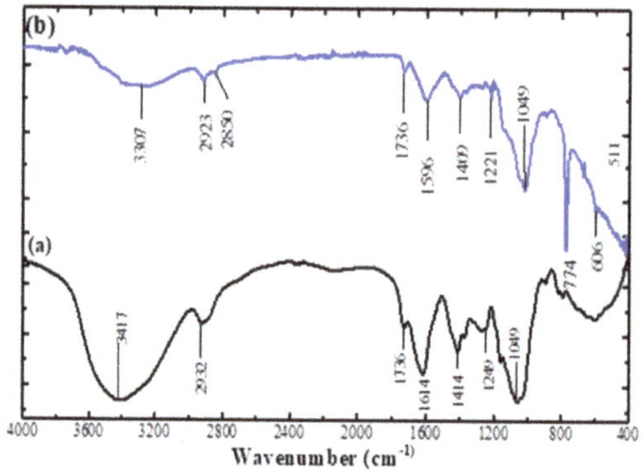

Figure 3. FTIR spectra of (**a**) xanthan gum (XG) (**b**) ZnO@XG nanocomposite.

3.2. XRD

The XRD spectra of bulk ZnO (black line) and ZnO@XG NPs (blue line) is given in Figure 4. The XRD of bulk ZnO NPs represented the characteristic peaks of ZnO NPs at 2θ values of 31.68°, 32.82°, 36.16°, 47.48°, 56.49°, 58.64°, 62.71°, and 67.81°, which correspond to miller indices values of (100), (002), (101), (102), (110), (103), (112), (200) crystalline plane of ZnO (JCPDS 89–0510) [43]. While looking at the spectra of ZnO@XG NPs, the characteristics peaks from ZnO NPs appeared at 2θ value 30.75° (ZnO), 33.63° (ZnO), 35.96° (ZnO) and 52.77° (ZnO) with corresponding miller indices values of (110), (002), (101), (200) and (202). The spectrum reveals the peaks with shifted diffraction angle (2θ values) from the precursor values with reduced intensity of ZnO due to functionalization with XG biopolymer chains, which imparted a small amorphous character to the nanoparticles.

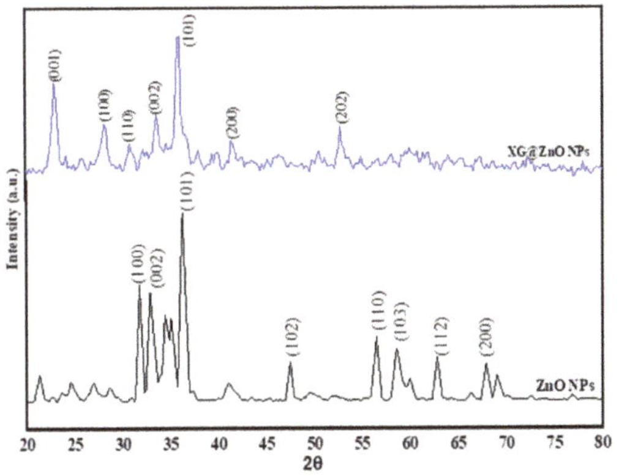

Figure 4. XRD spectra of ZnO (black line) and ZnO@XG (blue line).

Further information about the lattice structure, deformations on fusion and crystallite size can be obtained using Scherer's formula from Equations (1)–(4) (Scherrer, 1918, [44])

$$D = \frac{0.9\lambda}{\beta \cos\theta} \tag{1}$$

$$\text{DislocationDensity}(\delta) = \frac{1}{D^2} \tag{2}$$

$$\text{InterlayerSpacing}(d_{200}) = \frac{n\lambda}{2\text{Sin}\theta} \tag{3}$$

$$\%\text{Crystallinity} = \frac{\text{AreaUndertheCrystallinePeaks}}{\text{TotalArea}} \times 100 \tag{4}$$

where D is the crystal's size, λ is the wavelength used (i.e., 1.54 Å), β is the half-width of the most intense peak, and θ is the angle of diffraction. Using Equations (1)–(4), the average particle size of ZnO and ZnO@XG NPs was found to be 21.5 ± 1.5 and 14.7 ± 1.2 nm. The particle size of ZnO@XG NPs is also found to be in close agreement with the particle size (15.73 nm) obtained by TEM analysis. Hence, a decrease in particle size of ZnO NPs from 21.5 ± 1.5 to 14.7 ± 1.2 nm suggested the successful functionalization and reduction of Zn^{2+} ions to ZnO@XG NPs. The formation of the nanoparticles is also supported by the interlayer spacing value, which decreases from 0.24 Å in bare ZnO NPs to 0.18 Å in ZnO@XG NPs given in Table 1. These interactions of ZnO NPs and XG biopolymer chains resulted in a decreased value of dislocation density from 3.75×10^{15} to 2.95×10^{15} m^{-2} owing to contraction in size and change in crystallinity from 73% to 46% due to attachment of amorphous biopolymer chain. The XRD data analysis clearly suggested that there is a successful formation of ZnO NPs followed by surface functionalization by xanthan gum biopolymer chains.

Table 1. XRD parameters ZnO and ZnO@XG nanocomposite.

Component	2θ	FWHM (β_{hkl})	Interlayer Spacing (A°) at 2θ	Crystallite Size (nm) at 2θ	Dislocation Density (δ) × 10^{15} Lines (m^{-2})	% Crystallinity (%)
ZnO NPs	36.16	0.51	0.24	16.31	3.75	73
ZnO@XG	35.89	0.45	0.18	14.89	3.12	46

3.3. Morphological Analysis: SEM and TEM

Scanning electron microscopy (SEM) was employed to observe the surface morphological changes in the material during the solid-state reactions/interactions. Figure 5a,b represents the SEM image of ZnO@XG NPs at 7000×, 2 μm magnification ranges with EDX spectra within 1–20 KeV energy ranges. Figure 5a exhibits a highly porous surface morphology with a loosely agglomerated distribution of tiny nanowires of ZnO NPs on the surface (white dots) and black dots, represents the XG biopolymer matrix. Further, the atomic percentage of individual constituents used for the formation of ZnO@XG NPs was observed by the energy-dispersive X-rays (EDX) given in Figure 5b. The total output and conclusion received by EDX analysis express the composition of ZnO@XG NPs as C (72.16% ± 0.42%), O (26.42% ± 1.87%) and Zn (1.42% ± 0.48%). The transmission electron microscopy (TEM) was used for the elucidation of the optimized diameter and their variation in the XG biopolymer matrix. Figure 5c represents the TEM image of ZnO@XG NPs at 50 nm magnification range, which represents the loose agglomeration of tiny circular particles completely distributed along the XG matrix. The average size of nanorods was found to be 16.05 nm, which is in close concurrence with XRD and statistical Gaussian distribution analysis. Figure 5d was utilized to obtain the average particle size of ZnO NPs functionalized with XG biopolymer matrix using statistical domain tools like gaussian distribution. With a frequency of 16%, the average particle size was estimated as 15.73 nm, which is in close concurrence with XRD (14.7 nm) and TEM results (16.05 nm).

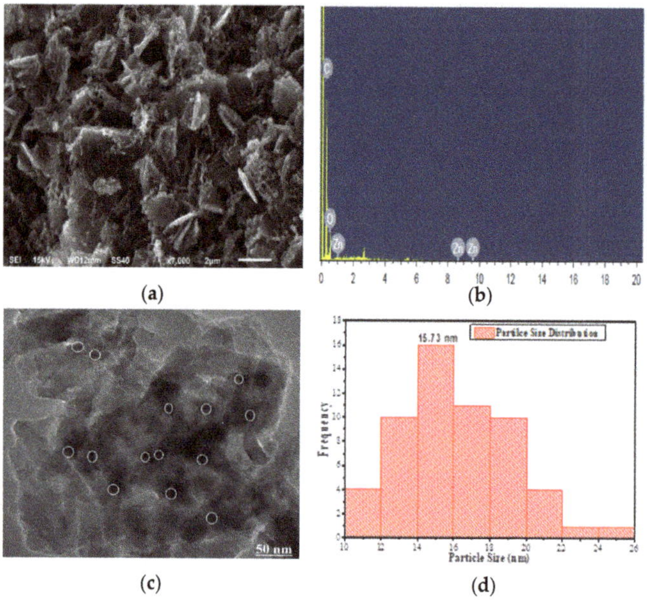

Figure 5. (**a**) SEM image of ZnO@XG nanocomposite (**b**) EDX spectra showing individual constituent elements comprising the material (**c**) TEM image of ZnO@XG nanocomposite showing the distribution of MSNs in the polymer matrix at 50 nm magnification range (**d**) Gaussian distribution of particle size for assessing the average particle size of nanomaterial.

3.4. BET

The Brunauer–Emmett–Teller (BET) isotherm for ZnO@XG NPs was acquired by nitrogen adsorption–desorption method. The BET plot for ZnO@XG NPs given in Figure 6 shows a type IV pattern, which suggested that the synthesized BNC has a nearly mesoporous structure [45]. The value of BET specific surface area for ZnO@XG NPs was found to be 9.24 $m^2 \cdot g^{-1}$ with a total pore volume of 0.045 $cm^3 \cdot g^{-1}$ and pore diameter of 12.87 nm. The reported values of BET specific surface area of bulk ZnO NPs synthesized by different routes are given as 15.45, 34.5, 12.998 and 7.5 $m^{-2} \cdot g^{-1}$ [42,46–48]. Hence, the reduction in specific surface area for the current BNC material suggests the incorporation of organic moieties of XG, which leads to blocking some pores due to surface functionalization.

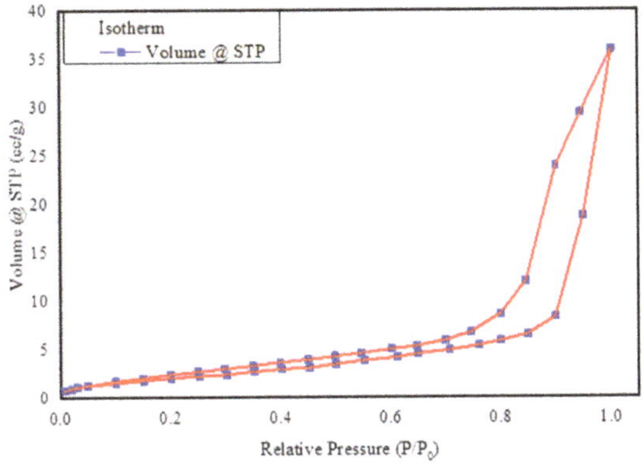

Figure 6. Low-temperature N_2 adsorption–desorption plot for ZnO@XG nanocomposite.

3.5. MIC

MIC of the synthesized ZnO@XG was determined against test pathogens *C. violaceum* and *S. marcescens*. MIC of ZnO@XG was recorded to be 256 µg/mL against both test pathogens. Since the current investigation was aimed to assess the quorum sensing and biofilm inhibitory of ZnO@XG, subinhibitory concentrations (0.0625–0.5xMICs) were selected for further microbiological assays.

3.6. QS Interference in C. violaceum

Sub-inhibitory concentrations (sub-MICs) of synthesized ZnO@Xanthan gum (ZnO@XG) nanocomposite were assessed for its quorum sensing (QS) inhibitory potential employing biosensor strain *C. violaceum* ATCC 12472 (CV12472). Production of violacein in *C. violacein* is regulated by the CviR-dependent quorum sensing system [49]. Impaired violacein production in CV12472 in a concentration-dependent manner is depicted in Figure 7B. Statistically significant ($p \leq 0.05$) reduction of 15%, 33%, 47% and 61% was over untreated control was recorded at 16, 32, 64, 128 µg/mL concentration of ZnO@XG nanocomposite, respectively. Tested sub-MICs (16–128 µg/mL) did not have any significant effect on the growth of the bacteria (Figure 7A), and thus, it is envisaged that the observed violacein reduction by nanocomposite is due to the quorum sensing interference rather than growth inhibition. Zinc oxide nanoparticles synthesized from plant extracts of *Nigella sativa* and *Ochradenus baccatus* have been reported with similar significant violacein inhibition activity (Al-Shabib et al., 2016, 2018, [50]). Chitinase production in *C. violaceum* is also QS regulated [29]. Chitinolytic activity in *C. violaceum* treated with sub-MICs of ZnO@XG was quantified using dye-release enzyme assay. The chitinolytic activity was reduced considerably with increasing concentration as compared to the control (Figure 7C). At 16, 32, 64, 128 µg/mL ZnO@XG treatment, 19%, 28, 54%, 70% reduced chitinolytic activity was observed, respectively. This is probably the first report demonstrating quorum sensing inhibition in *C. violaceum* by polysaccharide-based zinc oxide nanocomposite.

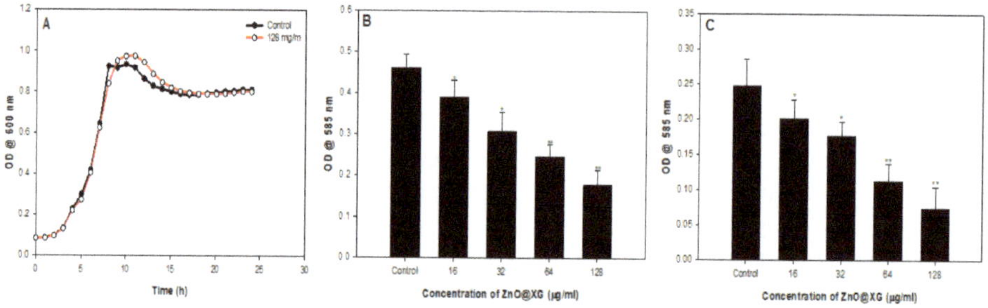

Figure 7. Effect of sub-MICs of ZnO@XG on (**A**) growth (**B**) violacein production and (**C**) chitinase production in *C. violaceum* ATCC 12,472. * denotes significance at $p \leq 0.05$, and ** denotes significance at $p \leq 0.005$.

3.7. QS Interference in *S. marcescens*

ZnO@XG was assessed for its QS inhibitory activity against virulence factors (prodigiosin and protease) produced by *S. marcescens*. Prodigiosin is a red pigment produced under the control of QS by *S. marcescens*. It is a vital virulence factor, playing a crucial role in the invasion, survival and pathogenicity of *S. marcescens* [51]. Figure 8B shows a concentration-dependent decrease in prodigiosin upon treatment with sub-MICs (16–128 µg/mL) of ZnO@XG. At 128 µg/mL concentration of ZnO@XG, prodigiosin declined by 71%, while at the lowest tested concentration (16 µg/mL) 25% decrease was recorded. Synthesized bio-nanocomposite did not affect the growth of the bacteria significantly at 128 µg/mL (Figure 8A). Our findings are in accordance with the results published on AgNPs synthesized from the extract of *Carum copticum*. At the highest tested concentration, AgNPs induced a 75% reduction in prodigiosin as compared to the untreated control [37].

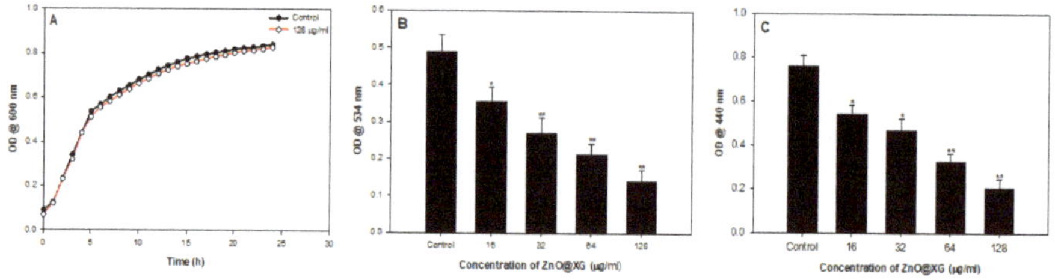

Figure 8. Effect of sub-MICs of ZnO@XG on (**A**) growth (**B**) prodigiosin production and (**C**) protease production in *S. marcescens* ATCC 13880. * denotes significance at $p \leq 0.05$, and ** denotes significance at $p \leq 0.005$.

Protease is another important QS regulated virulence factor produced by *S. marcescens*. Agents that can suppress the production of protease can be useful in potentiating the innate immune response of the host [52]. Therefore, we evaluated the effect of subinhibitory concentrations of ZnO@XG on protease production. The obtained results demonstrated that the production of protease decreased significantly ($p \leq 0.05$) at all tested concentrations (Figure 8C). Exposure to 128 µg/mL of ZnO@XG resulted in 72% less protease production as compared to the untreated control.

3.8. Effect on Biofilm and Biofilm-Related Virulence Functions

Considering the results of the virulence assays, we selected 64 (0.25xMIC) and 128 µg/mL (0.5xMIC) for further biofilm-related assays in *C. violaceum* and *S. marcescens*.

3.8.1. Inhibition of Biofilm Formation

The potential of the bacteria to cause infections and survive under stressed environments is often related to its capability to form biofilms. Biofilms are complex but organized structures of adherent bacteria forming microcolonies that are enveloped in a self-secreted matrix of EPS [53]. The role of QS in the regulation of various stages of biofilm formation like attachment and maturation is very well documented [54]. Biofilm inhibitory potential of ZnO@XG against both pathogens was evaluated using micro-titer plate (MTP) assay. ZnO@XG exposure caused a significant reduction in the biofilm-forming ability of both pathogens, viz. *C. violaceum* and *S. marcescens* (Figure 9A). At 64 μg/mL concentration, biofilm formation in *C. violaceum* and *S. marcescens* was inhibited by 49% and 53%, respectively, while at 128 μg/mL, it was further reduced by 67% (*C. violaceum*) and 77% (*S. marcescens*). Visual confirmation of the quantitative MTP biofilm inhibition assay was obtained using confocal laser scanning microscopic (CLSM) analysis (Figure 9B). CLSM image of the untreated control showed a closely-knit structure having dense aggregation of cells. ZnO@XG treatment resulted in considerably reduced biofilm formation marked by decreased surface coverage, scattered appearance of cells and disturbed integrity of biofilm (Figure 9B). Consistent with this report, Ravindran et al. (2018) reported significantly reduced biofilm formation in *S. marcescens* treated with AgNPs synthesized from the root extract of *Vetiveria zizanioides*.

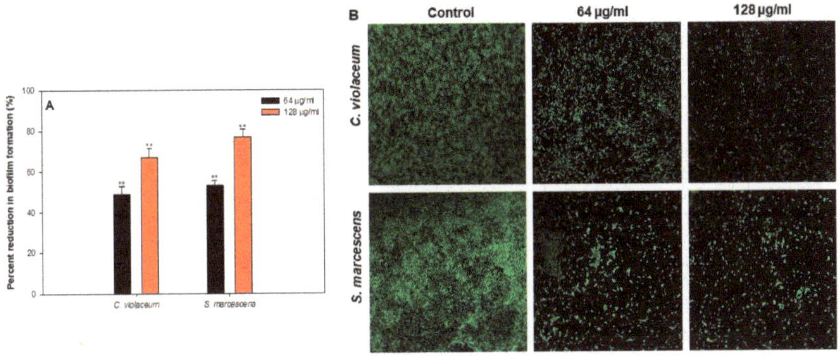

Figure 9. (**A**) Effect of sub-MICs of ZnO@XG on the biofilm formation of *C. violaceum* 12472, and *S. marcescens* ATCC 13880. Data are represented as mean values of triplicate readings, and the bar is the standard deviation. ** denotes significance at $p \leq 0.005$. (**B**) confocal laser scanning microscopic images of *C. violaceum* 12472 and *S. marcescens* ATCC 13880 biofilm in the absence and presence of sub-MIC ZnO@XG.

3.8.2. EPS Production

EPS is one of the most vital components of biofilm and plays a critical role in the attachment of cells to the substratum, maintenance of biofilm architecture, obtaining nutrients for cells, and protection of cells from the entry of antimicrobials [55]. EPS extracted from ZnO@XG treated and untreated cultures of *C. violaceum*, and *S. marcescens* was quantified, and the concentration-dependent decrease was recorded in both pathogens (Figure 10A). ZnO@XG at 128 μg/mL impaired the EPS production by 66% and 78% in *C. violaceum* and *S. marcescens*, respectively. Since ZnO@XG effectively reduces EPS production, it could possibly render the biofilm cells susceptible to the action of antibiotics. The observed results are in agreement with the findings of Hasan et al. (2019), [56], wherein dextrin-based poly(methyl methacrylate) grafted silver nanocomposites reduced EPS production significantly in drug-resistant bacteria and *Candida albicans*.

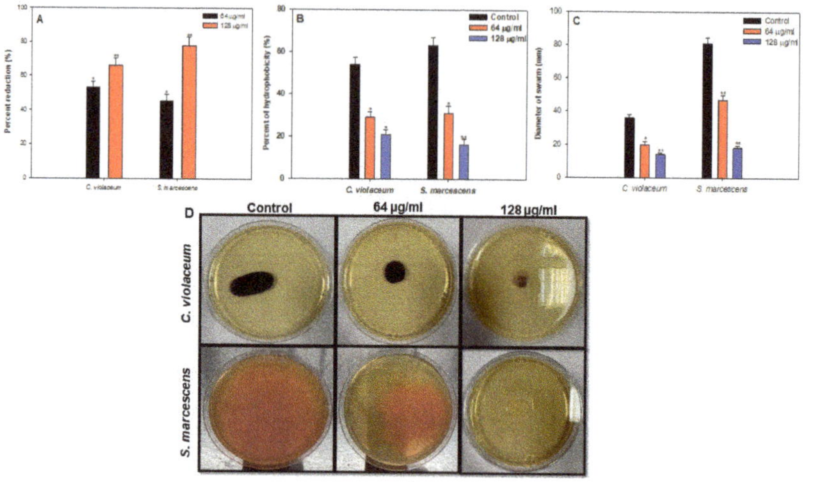

Figure 10. Effect of sub-MICs of ZnO@XG on (**A**) EPS production (**B**) cell surface hydrophobicity (**C**) swarming motility in *C. violaceum* 12472 and *S. marcescens* ATCC 13880. Data are represented as mean values of triplicate readings, and the bar is the standard deviation. * denotes significance at $p \leq 0.05$, and ** denotes significance at $p \leq 0.005$. (**D**) plates demonstrating swarming behavior of *C. violaceum* 12472 and *S. marcescens* ATCC 13880 in the absence and presence of sub-MIC ZnO@XG.

3.8.3. Cell-Surface Hydrophobicity (CSH)

Cell surface hydrophobicity is another important factor that contributes positively to the adhesion of microbial cells to the substratum. CSH facilitates adhesion by enhancing the hydrophobic interactions between the bacteria and biotic or abiotic surfaces [32]. MATH assay was employed to assess the effect of 64 and 128 µg/mL of ZnO@XG on CSH of the test pathogens. CSH of untreated controls of *C. violaceum* and *S. marcescens* was observed to be 54% and 63%, respectively (Figure 10B). CSH declined significantly with increasing concentration of ZnO@XG in both pathogens, and at 128 µg/mL, 21% and 16% CSH was demonstrated by *C. violaceum* and *S. marcescens* (Figure 10B). This drop in CSH upon treatment with sub-MICs of ZnO@XG could be responsible for the reduced biofilm formation by the test pathogens. In a study conducted on *P. aeruginosa* biofilm, it was envisaged that inhibition of CSH by copper nanoparticles was responsible for its biofilm inhibitory potential (LewisOscar et al., 2015, [57]).

3.8.4. Swarming Motility

Swarming motility is flagella-driven distinctive migration behavior in bacteria that plays a vital role in the inception of nosocomial infections. Furthermore, swarming motility is accountable for the enhanced biofilm formation of the pathogenic bacteria by facilitating the attachment of cells to the substratum [58]. Therefore, any interference in swarming behavior could lead to diminished biofilm formation. The diameter of the swarm of *C. violaceum* and *S. marcescens* was measured on 0.5% LB agar plates with or without sub-MICs of ZnO@XG. In *C. violaceum*, motility was reduced by 44% and 61% at 64 and 128 µg/mL concentrations, respectively (Figure 10C,D). Similarly, swarming was impaired by 42% and 77% in S. marcescens at 64 and 128 µg/mL concentrations, respectively (Figure 10C,D). Our findings corroborate well with a previous report demonstrating impaired swarming behavior in *P. aeruginosa*, *C. violaceum*, *S. marcescens* and *L. monocytogenes* upon treatment with 0.5xMICs of biologically synthesized tin oxide nanoflowers (Al-Shabib et al., 2018).

3.9. Disruption of Preformed Biofilm

Bacteria residing in the biofilm mode are many-fold more resistant to antibiotics, disinfectants and other bactericidal agents than their planktonic forms [59]. Therefore, disruption of preformed biofilms is rather difficult, and an effective biofilm inhibitory agent must be able to eradicate preformed mature biofilms. ZnO@XG was assessed for its ability to disrupt preformed mature biofilms at sub-MICs (0.25–0.5xMIC). Preformed biofilms were inhibited significantly at the tested concentrations in both pathogens (Figure 11). Observed results demonstrated that at 0.5xMIC (128 μg/mL) of ZnO@XG, preformed biofilms were eradicated by 54% in both *C. violaceum* and *S. marcescens*. EPS matrix envelopes the biofilm cells making them resistant to all kinds of bactericidal and bacteriostatic agents by blocking their entry. The obtained results demonstrate significant disruption of preformed biofilms upon treatment with ZnO@XG, indicating that the nanocomposite could breach the EPS matrix, disturb biofilm architecture and expose the bacterial cells rendering them susceptible to antimicrobials. This is probably the first report demonstrating inhibition of preformed biofilm of *C. violaceum* and *S. marcescens* by ZnO-biopolymer nanocomposite.

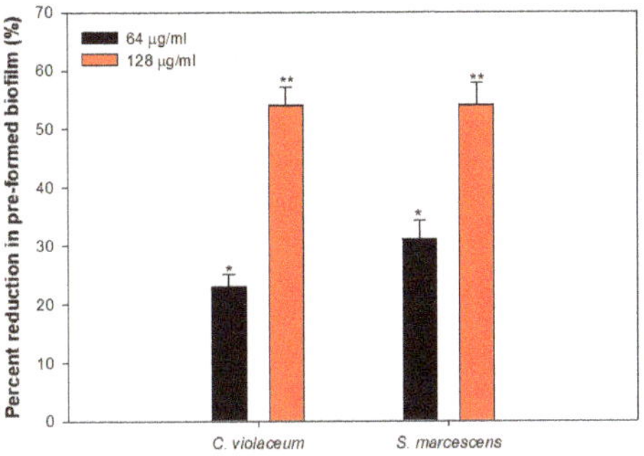

Figure 11. Effect of sub-MICs of ZnO@XG on preformed biofilm of *C. violaceum* 12472 and *S. marcescens* ATCC 13880. Data are represented as mean values of triplicate readings, and the bar is the standard deviation. * denotes significance at $p \leq 0.05$, and ** denotes significance at $p \leq 0.005$.

4. Conclusions

In conclusion, the study reports the successful formation of ZnO NPs followed by surface functionalization by xanthan gum biopolymer chains. Synthesized ZnO@XG were structurally and morphologically characterized using FTIR, XRD, BET, SEM and TEM. Subinhibitory concentrations of the biopolymer-based ZnO nanocomposite mitigated QS controlled virulence functions such as violacein, chitinase, prodigiosin, protease, EPS, swarming motility and CSH in pathogens, *C. violaceum* and *S. marcescens*. Further, ZnO@XG impaired biofilm formation and eradicated preformed biofilms, eventually leading to reduced pathogenicity of the test pathogens. Thus, it is envisaged that by targeting QS-regulated virulence, the likelihood of development of resistance is less, as no pressure is exerted on the growth of these pathogenic bacteria. Inhibition of virulence functions by ZnO@XG will disarm the bacteria so that they can be more easily eliminated by the host immune response. Thus, the synthesized ZnO@XG nanocomposites could be exploited to combat the threat of persistent bacterial infections and may also prevent the development of drug-resistance. Furthermore, the nanocomposite can be utilized in the food industry to prevent biofilm-based food contamination.

Author Contributions: Conceptualization, F.M.H. and I.H.; methodology, F.M.H., I.H., F.A.Q., R.A.K. and P.A.; software, I.H., R.A.K., and P.A.; validation, F.M.H., I.H., and A.A.; formal analysis, F.M.H. and I.H.; investigation, F.M.H., F.A.Q., and I.H.; resources, A.A.; data curation, F.M.H., F.A.Q., P.A., and I.H.; writing—original draft preparation, F.M.H., F.A.Q., and I.H.; writing—review and editing, F.M.H., F.A.Q., P.A., R.A.K., and I.H.; supervision, A.A.; project administration, A.A.; funding acquisition, A.A. All authors have read and agreed to the published version of the manuscript.

Funding: This research received no external funding.

Acknowledgments: The authors extend their appreciation to the Deputyship for Research and Innovation, "Ministry of Education" in Saudi Arabia, for funding this research work through project no. IFKSURG-1438-006.

Conflicts of Interest: The authors declare no conflict of interest.

References

1. Alizadeh-Sani, M.; Ehsani, A.; Moghaddas Kia, E.; Khezerlou, A. Microbial gums: Introducing a novel functional component of edible coatings and packaging. *Appl. Microbiol. Biotechnol.* **2019**, *103*, 6853–6866. [CrossRef] [PubMed]
2. Xu, W.; Jin, W.; Huang, K.; Huang, L.; Lou, Y.; Li, J.; Liu, X.; Li, B. Interfacial and emulsion stabilized behavior of lysozyme/xanthan gum nanoparticles. *Int. J. Biol. Macromol.* **2018**, *117*, 280–286. [CrossRef] [PubMed]
3. Ghorai, S.; Sarkar, A.; Panda, A.B.; Pal, S. Evaluation of the flocculation characteristics of polyacrylamide grafted xanthan gum/silica hybrid nanocomposite. *Ind. Eng. Chem. Res.* **2013**, *52*, 9731–9740. [CrossRef]
4. Joshy, K.S.; Jose, J.; Li, T.; Thomas, M.; Shankregowda, A.M.; Sreekumaran, S.; Kalarikkal, N.; Thomas, S. Application of novel zinc oxide reinforced xanthan gum hybrid system for edible coatings. *Int. J. Biol. Macromol.* **2020**, *151*, 806–813. [CrossRef]
5. Al-Shabib, N.A.; Husain, F.M.; Hassan, I.; Khan, M.S.; Ahmed, F.; Qais, F.A.; Oves, M.; Rahman, M.; Khan, R.A.; Khan, A.; et al. Biofabrication of zinc oxide nanoparticle from ochradenus baccatus leaves: Broad-spectrum antibiofilm activity, protein binding studies, and in vivo toxicity and Stress studies. *J. Nanomater.* **2018**, *2018*. [CrossRef]
6. Sirelkhatim, A.; Mahmud, S.; Seeni, A.; Kaus, N.H.M.; Ann, L.C.; Bakhori, S.K.M.; Hasan, H.; Mohamad, D. Review on zinc oxide nanoparticles: Antibacterial activity and toxicity mechanism. *Nano-Micro Lett.* **2015**, *7*, 219–242. [CrossRef]
7. Vikesland, P.; Garner, E.; Gupta, S.; Kang, S.; Maile-Moskowitz, A.; Zhu, N. Differential drivers of antimicrobial resistance across the world. *Acc. Chem. Res.* **2019**, *52*, 916–924. [CrossRef]
8. WHO. *The Top 10 Causes of Death*; WHO: Geneva, Switzerland, 2018.
9. Dadgostar, P. Antimicrobial resistance: Implications and costs. *Infect. Drug Resist.* **2019**, *12*, 3903–3910. [CrossRef]
10. WHO. *No Time to Wait: Securing the Future from Drug-Resistant Infections*; WHO: Geneva, Switzerland, 2019.
11. Shakoor, S.; Platts-Mills, J.A.; Hasan, R. Antibiotic-resistant enteric infections. *Infect. Dis. Clin. N. Am.* **2019**, *33*, 1105–1123. [CrossRef]
12. Andleeb, S.; Majid, M.; Sardar, S. Environmental and public health effects of antibiotics and AMR/ARGs. In *Antibiotics and Antimicrobial Resistance Genes in the Environment*; Elsevier: Amsterdam, The Netherlands, 2020; pp. 269–291.
13. Ahmad, I.; Qais, F.A.; Abulreesh, H.H.; Ahmad, S.; Rumbaugh, K.P. Antibacterial drug discovery: Perspective insights. In *Antibacterial Drug Discovery to Combat MDR*; Springer: Singapore, 2019; pp. 1–21.
14. Albrich, W.C.; Monnet, D.L.; Harbarth, S. Antibiotic selection pressure and resistance in streptococcus pneumoniae and streptococcus pyogenes. *Emerg. Infect. Dis.* **2004**, *10*, 514–517. [CrossRef]
15. Miller, K. Response of *Escherichia coli* hypermutators to selection pressure with antimicrobial agents from different classes. *J. Antimicrob. Chemother.* **2002**, *49*, 925–934. [CrossRef] [PubMed]
16. Rutherford, S.T.; Bassler, B.L. Bacterial quorum sensing: Its role in virulence and possibilities for its control. *Cold Spring Harb. Perspect. Med.* **2012**, *2*, a012427. [CrossRef] [PubMed]
17. Qais, F.A.; Khan, M.S.; Ahmad, I. Nanoparticles as quorum sensing inhibitor: Prospects and limitations. In *Biotechnological Applications of Quorum Sensing Inhibitors*; Kalia, V.C., Ed.; Springer: Singapore, 2018; pp. 227–244.
18. Davies, D.G.; Parsek, M.R.; Pearson, J.P.; Iglewski, B.H.; Costerton, J.W.; Greenberg, E.P. The involvement of cell-to-cell signals in the development of a bacterial biofilm. *Science* **1998**, *280*, 295–298. [CrossRef] [PubMed]
19. Percival, S.L.; Hill, K.E.; Williams, D.W.; Hooper, S.J.; Thomas, D.W.; Costerton, J.W. A review of the scientific evidence for biofilms in wounds. *Wound Repair Regen.* **2012**, *20*, 647–657. [CrossRef] [PubMed]
20. Martins, N.; Rodrigues, C.F. Biomaterial-related infections. *J. Clin. Med.* **2020**, *9*, 722. [CrossRef] [PubMed]

21. Lasa, I.; del Pozo, J.L.; Penadés, J.R.; Leiva, J. Biofilms bacterianos e infección. *An. Sist. Sanit. Navar.* **2005**, *28*, 163–175. [CrossRef]
22. Qais, F.A.; Ahmad, I. Green synthesis of metal nanoparticles: Characterization and their antibacterial efficacy. In *Antibacterial Drug Discovery to Combat MDR*; Springer: Singapore, 2019; pp. 635–680.
23. Al-Shabib, N.A.; Husain, F.M.; Ahmad, N.; Qais, F.A.; Khan, A.; Khan, A.; Khan, M.S.; Khan, J.M.; Shahzad, S.A.; Ahmad, I. Facile synthesis of tin oxide hollow nanoflowers interfering with quorum sensing-regulated functions and bacterial biofilms. *J. Nanomater.* **2018**, *2018*. [CrossRef]
24. Husain, F.M.; Ansari, A.A.; Khan, A.; Ahmad, N.; Albadri, A.; Albalawi, T.H. Mitigation of acyl-homoserine lactone (AHL) based bacterial quorum sensing, virulence functions, and biofilm formation by yttrium oxide core/shell nanospheres: Novel approach to combat drug resistance. *Sci. Rep.* **2019**, *9*, 1–10. [CrossRef]
25. Basha, S.K.; Lakshmi, K.V.; Kumari, V.S. Ammonia sensor and antibacterial activities of green zinc oxide nanoparticles. *Sens. Bio-Sens. Res.* **2016**, *10*, 34–40. [CrossRef]
26. Andrews, J.M. Determination of minimum inhibitory concentrations. *J. Antimicrob. Chemother.* **2001**, *48*, 5–16. [CrossRef]
27. Husain, F.M.; Ahmad, I. Doxycycline interferes with quorum sensing-mediated virulence factors and biofilm formation in Gram-negative bacteria. *World J. Microbiol. Biotechnol.* **2013**, *29*. [CrossRef] [PubMed]
28. Matz, C.; Deines, P.; Boenigk, J.; Arndt, H.; Eberl, L.; Kjelleberg, S.; Jurgens, K. Impact of violacein-producing bacteria on survival and feeding of bacterivorous nanoflagellates. *Appl. Environ. Microbiol.* **2004**, *70*, 1593–1599. [CrossRef] [PubMed]
29. Chernin, L.S.; Winson, M.K.; Thompson, J.M.; Haran, S.; Bycroft, B.W.; Chet, I.; Williams, P.; Stewart, G.S.A.B. Chitinolytic activity in chromobacterium violaceum: Substrate analysis and regulation by quorum sensing. *J. Bacteriol.* **1998**, *180*, 4435–4441. [CrossRef]
30. Champalal, L.; Kumar, U.S.; Krishnan, N.; Vaseeharan, B.; Mariappanadar, V.; Raman, P. Modulation of quorum sensing-controlled virulence factors in chromobacterium violaceum by selective amino acids. *FEMS Microbiol. Lett.* **2018**, *365*, 1–8. [CrossRef]
31. Morohoshi, T.; Shiono, T.; Takidouchi, K.; Kato, M.; Kato, N.; Kato, J.; Ikeda, T. Inhibition of quorum sensing in serratia marcescens AS-1 by synthetic analogs of N-acylhomoserine lactone. *Appl. Environ. Microbiol.* **2007**, *73*, 6339–6344. [CrossRef] [PubMed]
32. Ravindran, D.; Ramanathan, S.; Arunachalam, K.; Jeyaraj, G.P.; Shunmugiah, K.P.; Arumugam, V.R. Phytosynthesized silver nanoparticles as antiquorum sensing and antibiofilm agent against the nosocomial pathogen Serratia marcescens: An in vitro study. *J. Appl. Microbiol.* **2018**, *124*, 1425–1440. [CrossRef]
33. Al-Shabib, N.A.; Husain, F.M.; Nadeem, M.; Khan, M.S.; Al-Qurainy, F.; Alyousef, A.A.; Arshad, M.; Khan, A.; Khan, J.M.; Alam, P.; et al. Bio-inspired facile fabrication of silver nanoparticles from: In vitro grown shoots of Tamarix nilotica: Explication of its potential in impeding growth and biofilms of Listeria monocytogenes and assessment of wound healing ability. *RSC Adv.* **2020**, *10*. [CrossRef]
34. Al-Shabib, N.A.; Husain, F.M.; Qais, F.A.; Ahmad, N.; Khan, A.; Alyousef, A.A.; Arshad, M.; Noor, S.; Khan, J.M.; Alam, P.; et al. Phyto-mediated synthesis of porous titanium dioxide nanoparticles from withania somnifera root extract: Broad-spectrum attenuation of biofilm and cytotoxic properties against HepG2 cell lines. *Front. Microbiol.* **2020**, *11*, 1680. [CrossRef]
35. Husain, F.M.; Ahmad, I.; Baig, M.H.; Khan, M.S.; Khan, M.S.; Hassan, I.; Al-Shabib, N.A. Broad-spectrum inhibition of AHL-regulated virulence factors and biofilms by sub-inhibitory concentrations of ceftazidime. *RSC Adv.* **2016**, *6*. [CrossRef]
36. DuBois, M.; Gilles, K.A.; Hamilton, J.K.; Rebers, P.A.; Smith, F. Colorimetric method for determination of sugars and related substances. *Anal. Chem.* **1956**, *28*, 350–356. [CrossRef]
37. Qais, F.A.; Shafiq, A.; Ahmad, I.; Husain, F.M.; Khan, R.A.; Hassan, I. Green synthesis of silver nanoparticles using Carum copticum: Assessment of its quorum sensing and biofilm inhibitory potential against gram negative bacterial pathogens. *Microb. Pathog.* **2020**, *144*, 104172. [CrossRef] [PubMed]
38. Rosenberg, M.; Gutnick, D.; Rosenberg, E. Adherence of bacteria to hydrocarbons: A simple method for measuring cell-surface hydrophobicity. *FEMS Microbiol. Lett.* **1980**, *9*, 29–33. [CrossRef]

39. Al-Shabib, N.A.; Husain, F.M.; Rehman, M.T.; Alyousef, A.A.; Arshad, M.; Khan, A.; Masood Khan, J.; Alam, P.; Albalawi, T.A.; Shahzad, S.A.; et al. Food color 'Azorubine' interferes with quorum sensing regulated functions and obliterates biofilm formed by food associated bacteria: An in vitro and in silico approach. *Saudi J. Biol. Sci.* **2020**, *27*, 1080–1090. [CrossRef] [PubMed]
40. Mohsin, A.; Zhang, K.; Hu, J.; Tariq, M.; Zaman, W.Q.; Khan, I.M.; Zhuang, Y.; Guo, M. Optimized biosynthesis of xanthan via effective valorization of orange peels using response surface methodology: A kinetic model approach. *Carbohydr. Polym.* **2018**, *181*, 793–800. [CrossRef]
41. Chikkanna, M.M.; Neelagund, S.E.; Rajashekarappa, K.K. Green synthesis of Zinc oxide nanoparticles (ZnO NPs) and their biological activity. *SN Appl. Sci.* **2019**, *1*, 117. [CrossRef]
42. Thirumavalavan, M.; Huang, K.-L.; Lee, J.-F. Preparation and morphology studies of nano Zinc oxide obtained using native and modified chitosans. *Materials* **2013**, *6*, 4198–4212. [CrossRef]
43. Muhammad, W.; Ullah, N.; Haroon, M.; Abbasi, B.H. Optical, morphological and biological analysis of zinc oxide nanoparticles (ZnO NPs) using *Papaver somniferum* L. *RSC Adv.* **2019**, *9*, 29541–29548. [CrossRef]
44. Scherrer, P. Estimation of the size and internal structure of colloidal particles by means of X rays. *Nachs. Gissel. Wiss. Gott.* **1918**, *26*, 98–100.
45. Meng, A.; Xing, J.; Li, Z.; Li, Q. Cr-doped ZnO nanoparticles: Synthesis, characterization, adsorption property, and recyclability. *ACS Appl. Mater. Interfaces* **2015**, *7*, 27449–27457. [CrossRef]
46. Wang, J.; Xia, Y.; Dong, Y.; Chen, R.; Xiang, L.; Komarneni, S. Defect-rich ZnO nanosheets of high surface area as an efficient visible-light photocatalyst. *Appl. Catal. B Environ.* **2016**, *192*, 8–16. [CrossRef]
47. Le, T.K.; Nguyen, T.M.T.; Nguyen, H.T.P.; Nguyen, T.K.L.; Lund, T.; Nguyen, H.K.H.; Huynh, T.K.X. Enhanced photocatalytic activity of ZnO nanoparticles by surface modification with KF using thermal shock method. *Arab. J. Chem.* **2020**, *13*, 1032–1039. [CrossRef]
48. Nagaraju, P.; Puttaiah, S.H.; Wantala, K.; Shahmoradi, B. Preparation of modified ZnO nanoparticles for photocatalytic degradation of chlorobenzene. *Appl. Water Sci.* **2020**, *10*, 137. [CrossRef]
49. McLean, R.J.C.; Pierson, L.S.; Fuqua, C. A simple screening protocol for the identification of quorum signal antagonists. *J. Microbiol. Methods* **2004**, *58*, 351–360. [CrossRef] [PubMed]
50. Al-Shabib, N.A.; Husain, F.M.; Ahmed, F.; Khan, R.A.; Ahmad, I.; Alsharaeh, E.; Khan, M.S.; Hussain, A.; Rehman, M.T.; Yusuf, M.; et al. Biogenic synthesis of Zinc oxide nanostructures from Nigella sativa seed: Prospective role as food packaging material inhibiting broad-spectrum quorum sensing and biofilm. *Sci. Rep.* **2016**, *6*. [CrossRef]
51. Liu, G.Y.; Nizet, V. Color me bad: Microbial pigments as virulence factors. *Trends Microbiol.* **2009**, *17*, 406–413. [CrossRef]
52. Sethupathy, S.; Ananthi, S.; Selvaraj, A.; Shanmuganathan, B.; Vigneshwari, L.; Balamurugan, K.; Mahalingam, S.; Pandian, S.K. Vanillic acid from Actinidia deliciosa impedes virulence in Serratia marcescens by affecting S-layer, flagellin and fatty acid biosynthesis proteins. *Sci. Rep.* **2017**, *7*, 1–17. [CrossRef]
53. Hall-Stoodley, L.; Costerton, J.W.; Stoodley, P. Bacterial biofilms: From the natural environment to infectious diseases. *Nat. Rev. Microbiol.* **2004**, *2*, 95–108. [CrossRef]
54. Rice, S.A.; Koh, K.S.; Queck, S.Y.; Labbate, M.; Lam, K.W.; Kjelleberg, S. Biofilm formation and sloughing in serratia marcescens are controlled by quorum sensing and nutrient cues. *J. Bacteriol.* **2005**, *187*, 3477–3485. [CrossRef]
55. Zhou, J.-W.; Ruan, L.-Y.; Chen, H.-J.; Luo, H.-Z.; Jiang, H.; Wang, J.-S.; Jia, A.-Q. Inhibition of quorum sensing and virulence in serratia marcescens by hordenine. *J. Agric. Food Chem.* **2019**, *67*, 784–795. [CrossRef]
56. Hasan, I.; Qais, F.A.; Husain, F.M.; Khan, R.A.; Alsalme, A.; Alenazi, B.; Usman, M.; Jaafar, M.H.; Ahmad, I. Eco-friendly green synthesis of dextrin based poly (methyl methacrylate) grafted silver nanocomposites and their antibacterial and antibiofilm efficacy against multi-drug resistance pathogens. *J. Clean. Prod.* **2019**, *230*, 1148–1155. [CrossRef]
57. LewisOscar, F.; MubarakAli, D.; Nithya, C.; Priyanka, R.; Gopinath, V.; Alharbi, N.S.; Thajuddin, N. One pot synthesis and anti-biofilm potential of copper nanoparticles (CuNPs) against clinical strains of Pseudomonas aeruginosa. *Biofouling* **2015**, *31*, 379–391. [CrossRef] [PubMed]
58. Van Houdt, R.; Givskov, M.; Michiels, C.W. Quorum sensing in Serratia. *FEMS Microbiol. Rev.* **2007**, *31*, 407–424. [CrossRef] [PubMed]

59. Fux, C.A.; Costerton, J.W.; Stewart, P.S.; Stoodley, P. Survival strategies of infectious biofilms. *Trends Microbiol.* **2005**, *13*, 34–40. [CrossRef] [PubMed]

Publisher's Note: MDPI stays neutral with regard to jurisdictional claims in published maps and institutional affiliations.

coatings

MDPI

Article

Photochemical Preparation of Silver Colloids in Hydroxypropyl Methylcellulose for Antibacterial Materials with Controlled Release of Silver

Ondrej Kvitek [1,*], Elizaveta Mutylo [1], Barbora Vokata [2], Pavel Ulbrich [2], Dominik Fajstavr [1], Alena Reznickova [1] and Vaclav Svorcik [1]

1 Department of Solid State Engineering, University of Chemistry and Technology, Technicka 5, 166 28 Prague, Czech Republic; mutylol@vscht.cz (E.M.); fajstavd@vscht.cz (D.F.); reznicka@vscht.cz (A.R.); svorcikv@vscht.cz (V.S.)

2 Department of Biochemistry and Microbiology, University of Chemistry and Technology, Technicka 5, 166 28 Prague, Czech Republic; vokataa@vscht.cz (B.V.); ulbrichp@vscht.cz (P.U.)

* Correspondence: kviteko@vscht.cz; Tel.: +420-220-445-158

Received: 21 September 2020; Accepted: 27 October 2020; Published: 29 October 2020

Abstract: Silver nanoparticles (AgNPs) possess strong antibacterial effect. The current trend is to incorporate AgNPs into functional materials that benefit from their bactericidal capabilities. Hydroxypropyl methylcellulose (HPMC) is routinely used for the controlled release of medicine thanks to its slow dissolution in water and could be used as a matrix for the controlled release of AgNPs, if a method to produce such a material without the need of other reactants was developed. We proposed such a method in a photochemical reduction of $AgNO_3$ in hydroxypropyl methylcellulose (HPMC) solutions by the illumination of the mixture with the light emitting diode bulb for about 2 h. These AgNPs were characterized by transmission electron microscopy and their diameter was found to be mostly under 100 nm. The colloids were then easily transformed into solid samples by drying, lyophilization and spin-coating. The slowly soluble HPMC was found to be able to release the AgNPs gradually over the duration of several hours. Antibacterial activity of the prepared colloids and the solid samples was tested against *Escherichia coli* and *Staphylococcus epidermidis* and was found to be very high, reaching the total elimination of the bacteria in the studied systems.

Keywords: cellulose; silver; nanoparticle; antibacterial; composite; thin film

1. Introduction

Silver nanoparticles (AgNPs) are an intensively studied material due to their unique antibacterial properties combined with relatively low production costs. These particular properties predetermine the AgNPs for applications in areas such as functional textiles, active food packaging, medicine, cosmetics, ecology and many others [1–4]. In cosmetics, AgNPs have recently been used as additives for their antiseptic and preservative function in, e.g., acne treatment products [5]. Active food packaging is an innovative concept where the container interacts with the enclosed product and the surrounding environment to preserve the quality of the product during prolonged storage [6,7]. AgNPs supported on graphene oxide were infused into polyviscose pads to form antibacterial functional textile [8]. In medicine, AgNPs can be used in diagnosis (e.g., as plasmonic nano-antennas, for tunable wavelength imaging or surface-enhanced fluorescence [9]) or directly for treatment (e.g., as topical antimicrobial agents, wound dressing, implants or for cancer treatment) [10,11]. On the other hand, high concentrations of silver induce cytotoxicity and lead to argyria due to inappropriate silver accumulation in blood, lungs, liver and kidneys. The size of the AgNPs plays a great role in that smaller nanoparticles show both stronger antibacterial effects and are more toxic [10,12]. It is therefore necessary to control the size of the AgNPs during the preparation process (small enough particles to be effective in antimicrobial action, but big enough to be non-toxic) and to ensure the controlled release of the silver from the functional material to maintain safe but effective concentrations of the silver in the area of the medicament action [13].

The antibacterial activity of AgNPs is based on several interactions of the particles with the bacterial body. Very small AgNPs can penetrate the bacterial membrane and enter the cytoplasm where they induce strong oxidative stress. Larger nanoparticles accumulate on the bacterial membrane and form aggregates. This process deteriorates the integrity of the membrane which leads to cellular death. Both small and larger AgNPs release Ag^+ ions that bind to amine and thiol groups and thus disrupt the enzymatic respiratory system of the bacterial cell, destroy the DNA of the bacteria and break the proteins in the bacterial membrane [14–18].

AgNPs can be prepared by a variety of methods, but for colloids of monodisperse nanoparticles, the chemical reduction of Ag salts is the most common approach. The reaction mixture to prepare the AgNPs has to comprise of the Ag salt (most commonly $AgNO_3$ or AgCl), reducing agent to produce the Ag^0 and a stabilization agent to control the size of the nanoparticles as well as prevent their aggregation to larger structures [19,20]. The reducing and stabilization agents are often present in a great excess to assure the quick reaction, where a high number of metallic nuclei is created, which are immediately surface-stabilized and prevented from further growth [21]. Standard approaches utilize strong reducing agents such as $NaBH_4$, which are great for the highly controlled production of small monodisperse particles [22], however, a lot of attention was recently directed towards the use of less toxic alternatives, which are more desirable for medicinal purposes, due to the residues of the reducing and stabilizing agents being non-toxic [23,24]. Commonly used reactants include proteins such as collagen, gelatin or keratin [11] and polysaccharides such as chitosan [25–27] and cellulose [28–30]. Polysaccharides can also serve as stabilization agents, which enables the simple preparation of solid form nanoparticle composites by evaporating the solvent. Another way to eliminate the need for additional reaction agents is employing physicochemical methods of silver nitrate reduction. Because silver nitrate is very sensitive to visible light, it is possible to induce its reduction by the illumination of its solution in a photochemical reduction process [31,32]. It is still necessary for the reaction mixture to contain stabilizing agents to prevent the AgNPs from aggregating when the nitrate becomes reduced.

Cellulose by itself is insoluble and therefore cannot be easily used for nanoparticle production, but there are many soluble cellulose derivatives that are available for this purpose. Hydroxypropyl methylcellulose (HPMC) is a soluble cellulose derivative widely used in medicine for dosage forms, especially for controlled drug release tablets. It has been tested as a reducing and stabilizing agent for AgNPs as well [6,33,34]. The preparation of AgNPs directly in HPMC solution without additional reactants would represent an environmentally friendly and simple method to produce solid state composites of AgNPs for their controlled release which could be easily formed in desired shape.

In this work, the possibility to prepare AgNPs in HPMC solutions without additional chemicals was studied. Traditional wet chemistry ways to produce nanoparticles use excess reducing agents. The remaining substance would then crystallize after evaporation of the solvent, which hinders the preparation of solid materials (foils, coatings) with antibacterial activity from the colloids by drying, lyophilization or spin-coating. The reducing power of HPMC turned out to be insufficient to produce AgNPs from $AgNO_3$ by itself, hence photochemical reduction was chosen to produce the AgNPs in HPMC, because $AgNO_3$ is very sensitive to visible light. HPMC was chosen as a stabilizing agent due to its slow solubility in water and its capability to form a matrix for the AgNPs to be supported in. Thanks to those properties, HPMC is often used for the controlled release of drugs. The solid samples prepared as a combination of the AgNPs and HPMC could therefore be used for better control of AgNPs release from the functional materials, so the concentration of silver does not reach toxic levels.

2. Materials and Methods

2.1. Preparation of Ag Colloids

Silver colloids were prepared by photochemical method with hydroxypropyl methylcellulose (HPMC) serving as a capping agent. First, solutions of HPMC were prepared by the mixing of 12.5, 25 or 50 mg of HPMC powder (supplied by Merck (Darmstadt, Germany), viscosity 0.8–0.12 Pa·s in 2% water at 20 °C) and 45 mL of distilled water and the mixture was stirred in refrigerator overnight. Then, 5 mL of 35 mM solution of $AgNO_3$ (crystalline, supplied by Penta s.r.o., Prague, Czech Republic) was added dropwise to the HPMC solution at constant stirring in a shaded Erlenmeyer flask. The resulting mixture was therefore 3.5 mmol·L^{-1} $AgNO_3$ in 0.25%, 0.5% or 1% HPMC solution. Temperature was brought up to 60 °C and the solution was stirred further for 30 min. During this process, Ag seeds were created that catalyzed the photochemical reduction of the $AgNO_3$ in the next preparation step. The shade was then removed from the flask, the flask was moved to a closed box with a LED lamp light source (20 W power supply, 2452 lm luminous flux, emission spectrum measured by Red Tide USB650 spectrometer Ocean optics, Orlando, FL, USA) in Figure 1) in a 20 cm distance. The solution was irradiated for 2–6 h while its appearance changed from transparent colorless to dark yellow to reddish brown with apparent turbidity. The chemical reaction taking place can be written as: $2AgNO_3$ $2Ag + O_2 + 2NO_2$. The addition of 50 mg of glucose (D-glucose, supplied by Penta s.r.o.) as a reducing agent to the 50 mL of HPMC–$AgNO_3$ solution was tested as well.

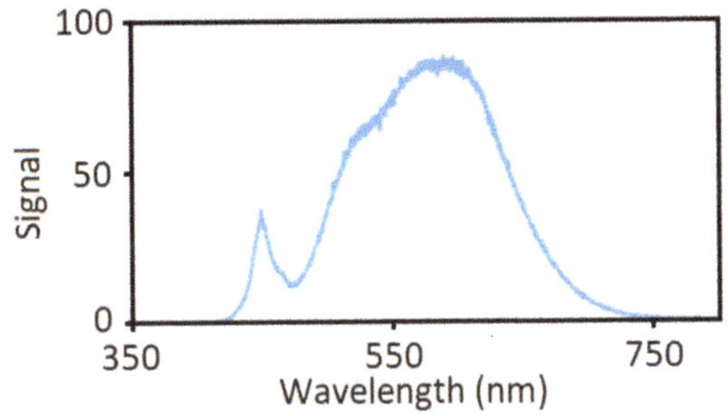

Figure 1. Normalized emission spectrum of an LED lamp employed as a light source for the photochemical reduction of $AgNO_3$ in hydroxypropyl methylcellulose (HPMC) solutions.

2.2. Preparation of Solid Samples

Solid samples of HPMC–AgNP composites were prepared from the colloids by three different methods. (i) Spin-coating method was used to prepare thin films on borosilicate glass slides (18 mm × 18 mm, 0.13–0.16 mm thick, supplied by Menzel-Gläser (Brunswick, Germany), thoroughly cleaned beforehand in methanol and dried in a stream of nitrogen gas) on Laurell WS-400B-6NPP spin-coater (North Wales, UK). The glass slides were spun at 1000, 1500 or 2000 rpm for 8 min after the injection of 1.5 mL of the colloid sample. (ii) Foils of the HPMC–AgNP composites were prepared by drying 4 mL of the colloid in round silicone forms, 5 cm in diameter, for several hours (until complete drying) at 70 °C in Binder FED 23 oven. (iii) Lastly, the sponges of the HPMC–AgNP composites were prepared by the lyophilization of 4 mL of the colloids in polystyrene test tubes. The samples were kept for 24 h in the freezer and then dried for 48 h in CentriVap lyophilizer (Labconco, Kansas City, MO, USA).

2.3. Methods of Analysis

Absorption spectra of the colloids and solid samples were measured by ultraviolet-visible-near-infrared (UV–Vis–NIR) spectrophotometer Lambda 25 (Perkin-Elmer, Waltham, MA, USA). The captured spectral range was in the range of 300–700 nm with a 1 nm data collection interval and a scanning rate of 240 nm·min^{-1}. Halogen and deuterium lamps were used as the light source for 300–350 and 350–700 nm, respectively.

Transmission electron microscopy (TEM) was employed to determine the shape and size of the prepared nanoparticles. The measurements took place in the JEOL JEM-1010 microscope (Tokyo, Japan) at the acceleration voltage of 80 kV. The AgNP colloids were applied to a copper mesh covered with carbon. The images were taken with SIS MegaView III (Olympus Soft Imaging Systems, Shinjuku, Japan) digital camera. The particle size distribution was evaluated with NIS-Elements AR software with 200–300 particles from each sample. The average size of the particles was obtained as a Feret diameter [35].

Visual characterization of the samples was performed by optical microscopy on confocal laser scanning microscope (CLSM) LEXT OLS 3100 (Olympus). Images of the samples were captured in TV mode using 5× and 10× objective lenses, which resulted in a total magnification of 120× and 240×, respectively.

A dual-beam scanning electron microscopy (SEM) with an field emission electron gun (Tescan LYRA3GMU, Brno, Czech Republic) was used to study the surface morphology of the solid samples. The applied acceleration voltage was 7 keV with 5 keV deceleration voltage. Prior to the measurement, the samples were attached to the sample holder with a carbon tape and coated with a thin layer of Au prepared by sputtering for 300 s at 40 mA in a Q300T ES (Quorum, Laughton, UK) sputter-coater to avoid charging.

UV–Vis spectroscopy (Lambda 25) was also employed to study the dissolution of the foils and sponges and the resulting release of the AgNPs from these solid samples. The samples were placed in a beaker with 20 mL of distilled water at 37 °C with constant stirring. The beaker was covered during the experiment to prevent the evaporation of the water. Then, 2 mL samples of the liquid part in the beaker were measured in cuvettes in predetermined intervals. After the measurement, the contents of the cuvette were returned into the beaker to ensure the constant volume of the solvent was maintained during the whole measurement. The experiment continued until the complete dissolution of the solid sample.

Gravimetry was employed to determine the thickness of the spin-coated thin films of the HPMC–AgNP composite. The cleaned glass slides were weighed before and after the thin film deposition at least 5 times each on the UMX2 (Mettler Toledo, Greifensee, Switzerland) ultra-microbalance system. The weight of the thin film (Δm) was calculated in film thickness (h) using the equation: $h = \Delta m/(a^2\ \rho)$, where a is the length of the glass slide side and ρ is the density of HPMC.

Fourier transform infrared (FTIR) spectroscopy was used to evaluate the changes of chemical composition in HPMC foils and in the thin films of HPMC with AgNPs prepared by the spin-coating method. Nicolet iS5 spectrometer (Thermo Fisher Scientific, Waltham, MA, USA) was used for this purpose with iD7 attenuated total reflection accessory with the diamond crystal. The spectral range was 600–4000 cm^{-1} with 1 cm^{-1} data interval and the spectra were obtained as an average from 128 measurement cycles. Automatic atmospheric suppression utility was used to filter out the absorption of ambient CO_2 and H_2O.

Atomic force microscopy (AFM) was performed on the spin-coated thin films of HPMC to investigate the surface morphology of the prepared samples. Dimension ICON (Bruker, Billerica, MA, USA) microscope was used in tapping mode with Si ScanAsyst probe with 70 kHz resonance frequency and 0.4 N·m^{-1} spring constant. Images of 10 μm × 10 μm were obtained at a 0.3 Hz scanning rate.

2.4. Antibacterial Activity Tests

Antibacterial activity of the prepared samples was tested against the Gram-negative *Escherichia coli* (*E. coli*; DBM 3138) bacteria and Gram-positive *Staphylococcus epidermidis* (*S. epidermidis*; DBM 3179) bacteria. The bacteria were cultivated overnight on sterile Luria–Bertani growth medium (LB, Merck, prepared by dissolving 10 g of tryptone, 5 g of yeast extract and 5 g of NaCl in 1 L of distilled water) with constant shaking at 37 °C. The suspension with bacteria was then diluted with fresh sterile phosphate-buffered saline (PBS, Merck, prepared by dissolving 1 tablet in 200 mL of distilled water) to a concentration of 10^3–10^4 bacteria per 1 mL. HPMC–AgNP colloids were tested by adding 100 µL of the colloid to 2 mL of the bacterial suspension. The mix was incubated for 2–3 h at room temperature and then 5 µL × 25 µL of each triplicate sampling was placed on a Petri dish. The spin-coated HPMC–AgNP composite thin films were tested by placing 150 µL of the bacterial suspension on the solid sample in 5 repetitions. After 2–3 h of incubation, 3 × 25 µL of the bacterial suspension was placed on a Petri dish. All the bacterial suspensions on Petri dishes were then incubated for 24 h with LB at room temperature for *E. coli* and with plate-count agar at 37 °C for *S. epidermidis*. The bacterial concentration was determined by directly counting the colony forming units on the Petri dishes.

3. Results

The colloids were prepared by a method similar to the one published in [33] with glucose as a supportive reducing agent. However, the glucose excess impedes the preparation of solid samples as it crystallizes in the foils as the solvent dries (Figure 2). The glucose was omitted from the reaction mixture and the procedure described in [34] was adopted. With the reaction taking place in a shaded flask, no color change indicating the formation of AgNPs occurred. When the samples were then left in the daylight in the laboratory the reduction of $AgNO_3$ occurred. $AgNO_3$ is very sensitive to illumination and therefore the preparation sequence was modified to include the controlled light irradiation of the colloid samples, as described in the experimental section. During the irradiation, a gradual color change took place from clear solutions to reddish brown colloids after 2 h of irradiation.

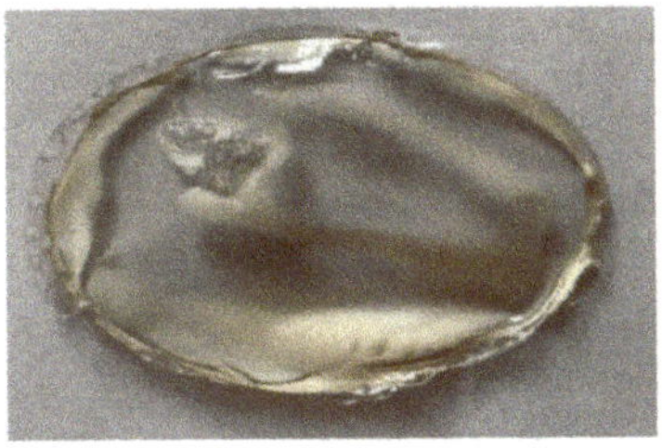

Figure 2. Photograph of a dried foil prepared from the HPMC mixed with glucose as a reducing agent. White opaque streaks of crystallized glucose are clearly visible in the otherwise transparent HPMC foil.

The UV–Vis spectra of Ag colloids prepared in 0.25%, 0.5% and 1% solutions of HPMC are shown in Figure 3. The gradual increase in the absorption of the surface plasmon resonance (SPR) band in the spectra with increasing irradiation time is visible. The concentration of HPMC in the solutions influences the overall absorption of the samples as well. The absorption maximum of the SPR band is at 470 to 490 nm with a shoulder occurring in the case of longer irradiation times at wavelengths around 600 nm, which can be attributed to quadrupole and dipole SPR oscillations (preferential shape growth or the presence of two distinct populations of nanoparticles of different sizes were disproven by TEM measurements) [36]. This suggests the formation of somewhat larger AgNPs of about 150 nm. The interband transition absorption band can be seen at all HPMC concentrations at 380 nm and is more prominent in samples with higher HPMC concentration.

The TEM images of the AgNPs prepared in 0.25%, 0.5% and 1% HPMC solutions are shown in Figure 4a–c. Well defined metal particles are clearly visible documenting the successful reduction of $AgNO_3$. With higher magnification, crystalline facets can be distinguished as well. The size distribution of the particles is obviously quite wide as there is a small number of significantly larger particles in the images. The particles mainly possess a round shape with no significant elongation in any axis. The size distribution seems to be in a spectrum of diameters rather than several distinct populations of different particle sizes. The image analysis of the TEM micrographs was used to quantify the size distribution of the particles as well (Figure 4d). The narrowest size distribution was found to be in the case of the 0.25% HPMC solution, where the majority of the particles was found to be 60–120 nm in diameter. That is still a rather wide size distribution, but the conditions of the proposed preparation method could be adjusted to achieve better results in future experiments. The highest population of the smallest AgNPs was found to be in the case of the 1% HPMC solution where most of the particles were found to be under 50 nm in diameter, but there were still many particles significantly bigger. The sample with 0.5% HPMC solution showed the widest size distribution of the AgNPs and showed the characters of size distributions from the other two studied HPMC concentrations with the highest number of particles being 10–40 and 80–110 nm in diameter. The wide size distribution is undesirable for the controlled antibacterial effect of the colloid.

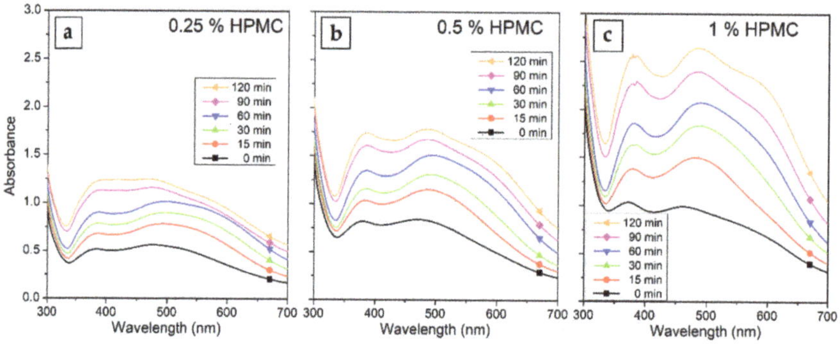

Figure 3. UV–Vis absorption spectra of Silver nanoparticles (AgNPs) colloids prepared by the irradiation of $AgNO_3$ solutions in (**a**) 0.25%, (**b**) 0.5% and (**c**) 1% HPMC for 15–120 min.

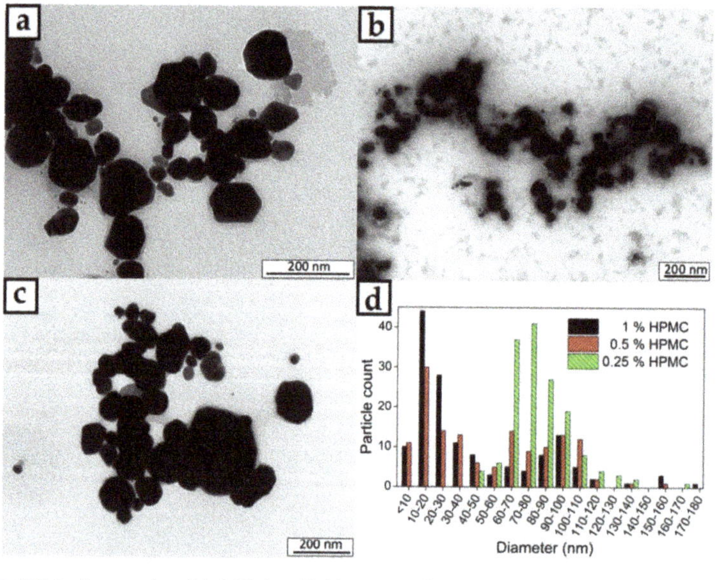

Figure 4. TEM micrographs of AgNPs in colloids prepared in (**a**) 0.25%, (**b**) 0.5% and (**c**) 1% HPMC solutions; (**d**) graph of particle size distributions of the prepared colloid samples.

The colloids were then dried in an oven in silicon molds, where the HPMC–AgNP composite foils formed. A photograph of the prepared foil can be seen in Figure 5a. The foil showed good mechanical properties, it was flexible and did not break when bent. The color of the foils was dark yellow-brown with no significant AgNPs aggregates that could be spotted by the naked eye. The micrographs of the foil surface obtained from CLSM (Figure 5b) and SEM (Figure 5c,d) show the homogeneous flat surface of the foils. No significant surface irregularities can be spotted even under the 2000× magnification of the SEM image. The higher magnification SEM image also show the homogeneous distribution of AgNPs in the HPMC foil. The HPMC solutions produced homogeneous foils with smooth surface. The HPMC–AgNP colloids were also lyophilized to produce dry sponges of the composite material. A photograph and micrographs of the sponge are in Figure 6. The color of the sponge was similar to that of the dry foil with no apparent AgNP aggregates. The colloids are therefore stable enough to endure the lyophilization process. The micrographs show a structure of very fine leaflets of the cellulose with an apparently much higher surface area than that of the dry foils. The prepared sponges were flexible and able to regain their initial shape after being mechanically deformed.

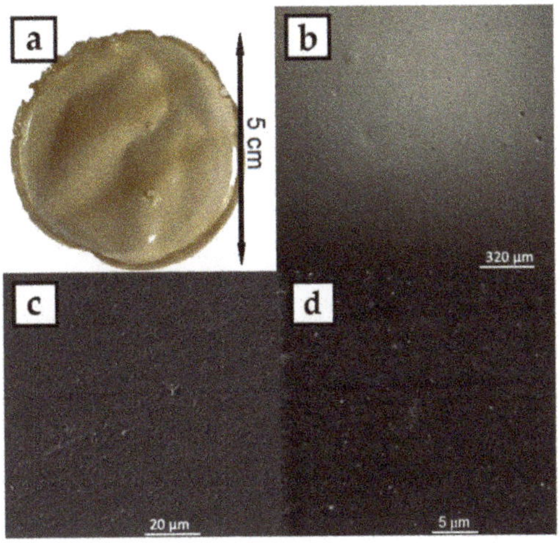

Figure 5. (**a**) Photograph, (**b**) CLSM and (**c**,**d**) SEM micrographs of AgNPs in HPMC composite foil prepared by drying the 1% HPMC colloids.

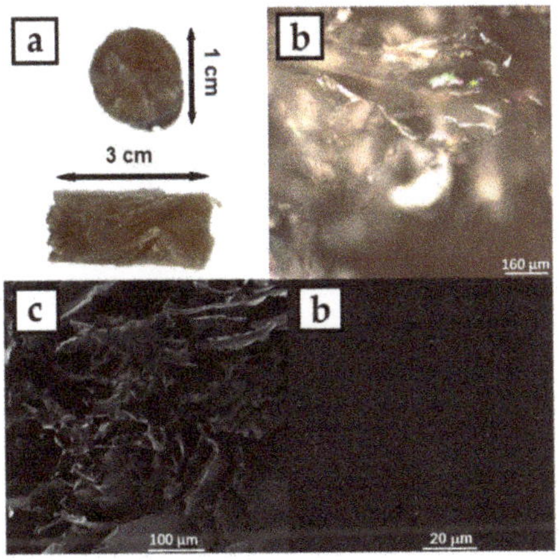

Figure 6. (**a**) Photograph (side and front), (**b**) CLSM and (**c**,**d**) SEM micrographs of AgNPs in HPMC composite sponge prepared by lyophilizing the 1% HPMC colloids.

FTIR absorption spectra in Figure 7a show no significant difference of the dry foil with or without the AgNPs. This could be expected, because there is no covalent bond between the HPMC and Ag and the nanoparticles are stabilized only sterically. The spectra show strong absorption at 1050 cm^{-1} which can be attributed to C–O–C stretching vibration, and 1380 and 3460 cm^{-1} absorption can be attributed to –OH bending and stretching vibrations, the 1460 cm^{-1} band represents the –CH_2-scissor bend vibration and lastly the absorption at 2930 cm^{-1} is connected to the C–H stretching vibrations. The foil sample containing the AgNPs only shows the comparatively weaker absorption of the 1050 cm^{-1} band. This could be caused by the interaction of this chemical group with the surface of the AgNPs. FTIR spectra of the thin film samples in Figure 7b show similar absorptions as the dry foils at 3460 and 2930 cm^{-1}. The absorption bands of the cellulose at lower wavenumbers are however overlapped by the much stronger absorption of the glass substrate, particularly the BO_3 triangles stretching vibration at 1390 cm^{-1} and the SiO_4 tetrahedral units stretching vibration at 1100 cm^{-1}.

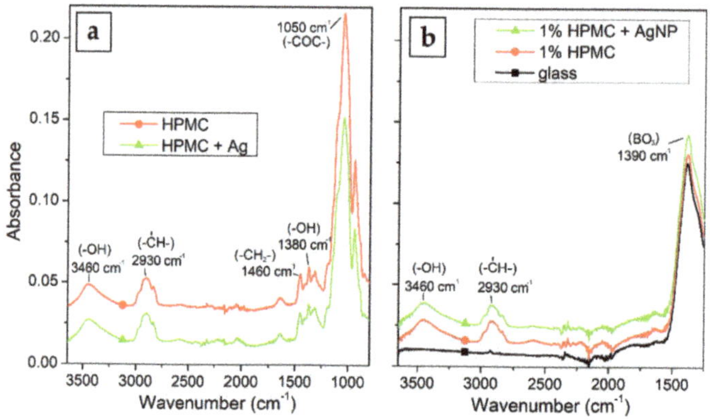

Figure 7. FTIR absorption spectra of (**a**) HPMC foil and HPMC–AgNP composite foil prepared from 1% HPMC solutions and (**b**) borosilicate glass substrate and thin films of HPMC and AgNPs prepared by spin-coating.

The HPMC is characteristically better soluble in water at lower temperatures (best at 4 °C) [37]. Therefore, the controlled dissolution of the solid samples prepared from the 1% HPMC colloids and slow release of AgNPs was tested at the physiological temperature of 37 °C. The UV–Vis absorption of the solutes taken at different times are in Figure 8. The foil dissolved overall much more quickly than the sponge. The foil got completely dissolved after 2 h. The dissolution was quite fast at first, after about 1 h most of the sample got dissolved. The rest of the sample took another hour to become completely dissolved. After up to 1 h of the process, a significantly sharper SPR band than that of the original colloid at around 420 nm can be seen in the spectrum of the solute. This could suggest that smaller AgNPs are released preferentially from the foil. On the other hand, the composite sponge dissolved much slower, the complete dissolution was observed at about 20 h after the start of the process. The SPR band in this case remains wide during the whole dissolution process, because the sponge dissolves in whole segments, where no preferential release of smaller AgNPs could take place.

The prepared AgNP colloids in HPMC solution were also used to form thin films on glass slide substrates by spin-coating. The dependence of the prepared film thickness on HPMC concentration and rotation speed of the spin-coater calculated from the weight difference of the samples is in Figure 9. The film is obviously thicker at higher HPMC concentrations and lower rotation speeds ranging from about 180 nm at 1% HPMC and 1000 rpm to about 20 nm at 0.25% HPMC and 2000 rpm. The UV–Vis absorption spectra of the prepared composite thin films in Figure 10 shows rather than low values of absorption due to the low thickness of the prepared films. The overall absorption is not in all cases higher for the thicker prepared films, which is probably caused by inhomogeneities in the prepared thin films (e.g., in the case of the 0.5% HPMC solution, we expect the film prepared at 1000 rpm to have the highest absorbance). However, the absorption of the thin films was so low that these results could also be influenced by the uneven thickness of the glass substrates. In the case of the thin films prepared from the 0.25% and 0.5% solutions, a single SPR band is present in the spectra at about 450 and 480 nm, respectively. The spectra of the samples prepared from the 1% solutions are, on the other hand, similar to those of the colloids with the quadrupole absorption peak present. In the case of the 1% 1000 rpm sample, a strong inter-band absorption at 380 nm is present as well, which could be related to the presence of Ag agglomerates in the thin film.

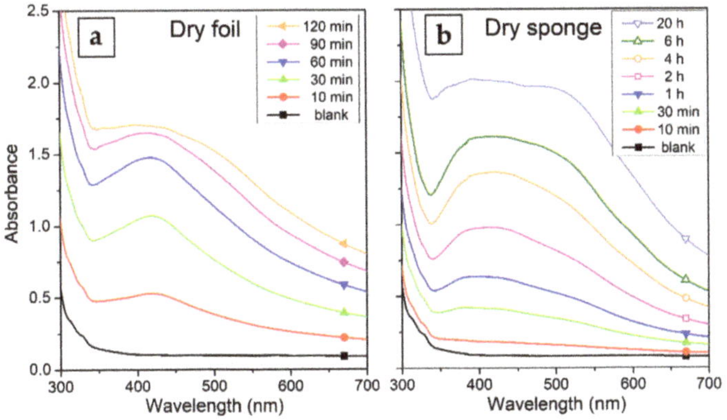

Figure 8. UV–Vis absorption spectra of the solutes taken during the dissolution of the HPMC–AgNP composite: (**a**) dry foil (solutes sampled after 10–120 min dissolution) and (**b**) lyophilized sponge (solutes sampled after 10 min to 20 h of dissolution).

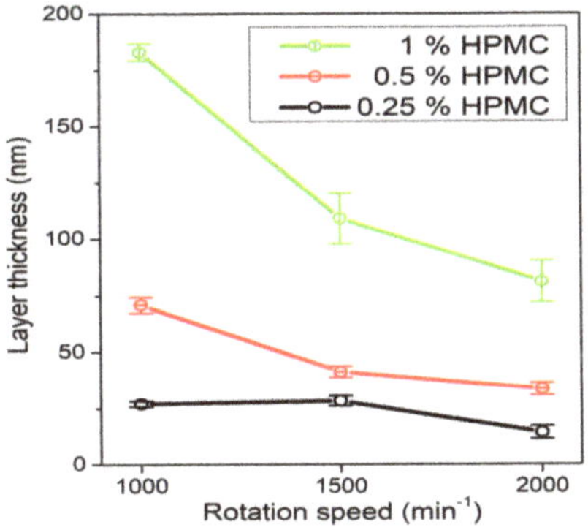

Figure 9. Dependence of spin-coated film thickness on the rotation speed of the spin-coater table for the samples prepared from 0.25%, 0.5% and 1% HPMC colloids.

The surface morphology of the spin-coated thin films on glass slides was studied by AFM (Figure 11). The films prepared from HPMC solutions without the AgNPs show a very homogeneous and smooth surface. This shows the capability of the HPMC to form delicate thin films from solutions at the chosen preparation conditions. In the case of the films with AgNPs, the top parts of the particles can be seen sticking from the film's smooth surface in the micrographs. The thicker films prepared from higher HPMC (particularly the 1% HPMC–AgNP sample) concentrations are able to support larger particles that rise up to about 150 nm above the film surface. Many smaller particles can be seen on the images as well, scattered on the film's surface. With the lower thickness of the films the larger particles could not be captured in the volume of the film, therefore a higher number of smaller particles are sticking from the surface, while the larger particles were washed away from the surface during the spin-coating.

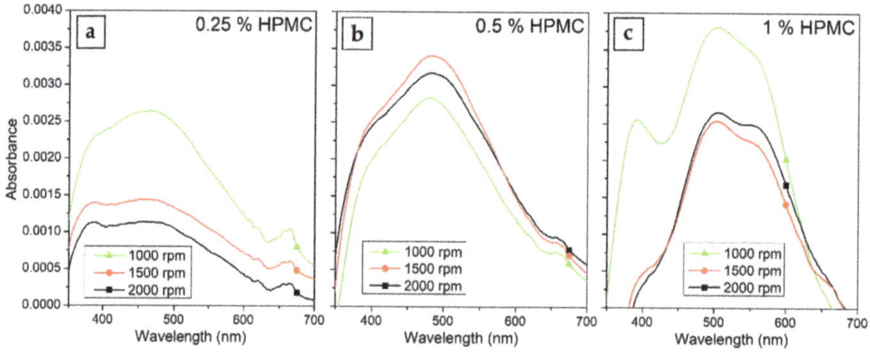

Figure 10. UV–Vis absorption spectra of the composite HPMC–AgNP spin-coated thin films prepared from (**a**) 0.25%, (**b**) 0.5% and (**c**) 1% HPMC colloids at 1000–2000 rpm at spin-coater table rotation speed.

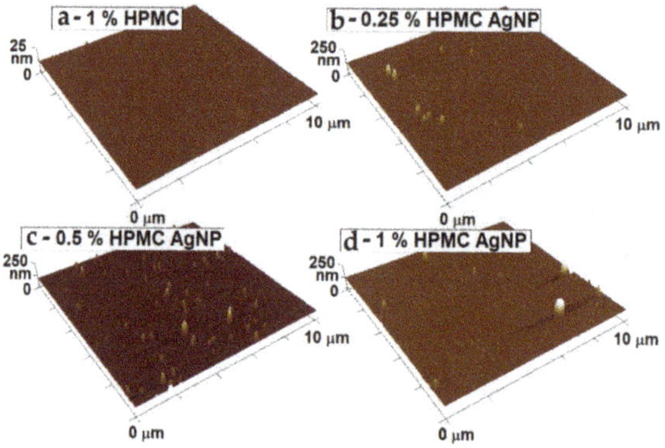

Figure 11. Surface morphology of the blank HPMC and HPMC with AgNP thin films measured by atomic force microscopy (AFM). Notice the different z axis scale in the case of the blank HPMC sample. The thin films were prepared from: (**a**) 1% solution of pure HPMC; samples of AgNPs prepared in (**b**) 0.25%; (**c**) 0.5% and (**d**) 1% HPMC solutions.

The thin film solid samples prepared at 1000 rpm were also used to test their antibacterial activity against Gram-negative *E. coli* and Gram-positive *S. epidermidis* bacterial cells. The dispersed bacteria were kept in contact with the solid samples or colloids and cultivated afterwards. A 1% HPMC solution without the AgNPs was used to prepare the thin film for the blank test. In the case of the blank sample of clean HPMC, no significant decrease in the number of bacteria was observed in most cases, except *S. epidermidis* against the thin HPMC film, where the number of bacterial colonies was reduced to about one third compared with the control. All the samples with AgNPs were then able to eliminate all the bacteria from their dispersions and no bacterial colonies could then be cultivated from both the dispersions that were in contact with the solid samples or HPMC–AgNP colloids (Figure 12).

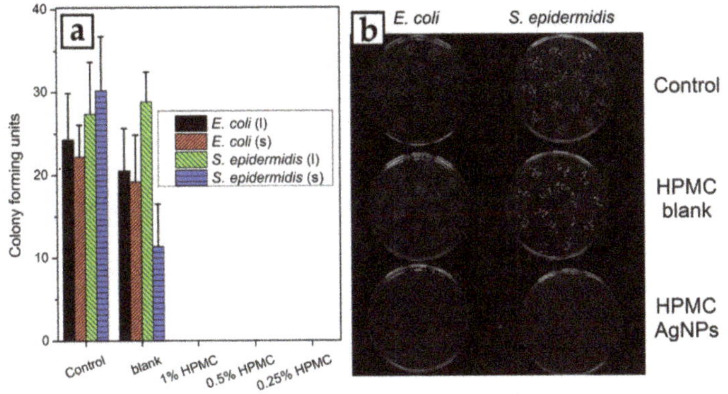

Figure 12. (**a**) Results of the antibacterial activity tests of the HPMC–AgNP colloids (columns marked l) and spin-coated thin films of HPMC–AgNP composites (columns marked s) against *E. coli* and *S. epidermidis* bacterial cells. (**b**) Images of Petri dishes after bacterial cultivation in contact with thin-film samples prepared from pure 1% HPMC solution (HPMC blank) and from 1% HPMC solution with AgNPs (HPMC–AgNPs).

4. Discussion

AgNP colloids were prepared in HPMC solutions by photochemical reduction method without the presence of additional reducing or stabilizing agents. The reducing power of HPMC by itself was found to be insufficient (contrary to what was reported in [34]) to produce AgNPs from $AgNO_3$. The prepared colloid dispersions were used for the simple preparation of solid composite samples by drying, lyophilization and spin-coating methods. That would not be possible if additional reducing agent was added, because the excess agent would crystallize during the drying of the colloid. Therefore, the photochemical reduction method was found to be most suitable to produce colloids that can be easily formed into solid material.

The structure, size and polydispersity of the prepared AgNPs was studied by TEM, which showed the rather wide size distribution of the prepared AgNPs. Most of the particles were found to be smaller than 100 nm in diameter. This has an effect on other parameters of the colloids, and the UV–Vis spectra showed broad SPR absorbances with dipole and quadrupole oscillations. The proposed preparation procedure and conditions should be therefore further explored to improve on the size distribution and control of the AgNP size, since in comparison to photochemically reduced AgNPs in polyvinyl pyrrolidone [31,32], the size distribution of the AgNPs was rather wide.

On the other hand, HPMC showed to be advantageous in terms of solid sample preparation. The composite foils prepared by drying at 70 °C showed good mechanical cohesion and elasticity with no tendency to break apart during manipulation with the samples. Sponges prepared by lyophilization showed a leaflet-like structure and were mechanically resilient as well. Dissolution of the foils in water showed the preferential release of smaller AgNPs with the UV–Vis spectra of the solutes showing much narrower SPR absorption than that of the original colloid. The dissolution of HPMC was gradual over the course of several hours to about 1 day, depending on the preparation method of the solid sample. This result is very promising in the sense of the controlled release of AgNPs from the functional material. The colloids were also very suitable for the preparation of thin films on glass substrates by the spin-coating method. The samples prepared in this way showed a very strong antibacterial effect against *E. coli* and *S. epidermidis*.

The prepared materials show promise for a new branch of antibacterial materials based on AgNPs. The controlled release of AgNPs from the material would take course over several hours to days and the concentration of the silver in the vicinity of the material could be controlled by the dissolution rate of the HPMC composite. Further development of the proposed material is, however, needed. The employed concentrations of AgNPs were able to eliminate bacteria in the tests easily, and the concentration could therefore be tuned to lower levels of silver content. The size dispersion of the prepared nanoparticles turned out to be rather wide and further improvements to the preparation method could be explored (e.g., better control of the seeding process, modification of the reactant concentrations or tuning the wavelength, heating and stirring during the photoreduction process). Lastly, cytotoxicity tests would need to be performed to determine the high end of the desired concentration of these particular AgNPs in the antibacterial materials.

Author Contributions: Conceptualization, O.K. and A.R.; methodology, O.K. and E.M.; validation, O.K.; formal analysis, O.K. and E.M.; investigation, E.M., B.V., P.U. and D.F.; resources, O.K. and A.R.; data curation, E.M.; writing—original draft preparation, O.K.; writing—review and editing, A.R. and V.S.; visualization, O.K.; supervision, O.K. and V.S.; project administration, O.K.; funding acquisition, V.S. All authors have read and agreed to the published version of the manuscript.

Funding: This research was funded by Czech Science Foundation (GA CR) under the project No. 20-01641S.

Conflicts of Interest: The authors declare no conflict of interest. The funders had no role in the design of the study; in the collection, analyses, or interpretation of data; in the writing of the manuscript, or in the decision to publish the results.

References

1. Keat, C.L.; Aziz, A.; Eid, A.M.; Elmarzugi, N.A. Biosynthesis of nanoparticles and silver nanoparticles. *Bioresour. Bioprocess.* **2015**, *2*, 1–11. [CrossRef]
2. Hebeish, A.; El-Naggar, M.E.; Fouda, M.M.G.; Ramadan, M.A.; Al-Deyab, S.S.; El-Rafie, M.H. Highly effective antibacterial textiles containing green synthesized silver nanoparticles. *Carbohydr. Polym.* **2011**, *86*, 936–940. [CrossRef]
3. Tran, Q.H.; Nguyen, V.Q.; Le, A.T. Silver nanoparticles: Synthesis, properties, toxicology, applications and perspectives. *Adv. Nat. Sci. Nanosci.* **2013**, *4*, 033001. [CrossRef]
4. Morley, K.S.; Webb, P.B.; Tokareva, N.V.; Krasnov, A.P.; Popov, V.K.; Zhang, J.; Roberts, C.J.; Howdle, S.M. Synthesis and characterisation of advanced UHMWPE/silver nanocomposites for biomedical applications. *Eur. Polym. J.* **2007**, *43*, 307–314. [CrossRef]
5. Kokura, S.; Handa, O.; Takagi, T.; Ishikawa, T.; Naito, Y.; Yoshikawa, T. Silver nanoparticles as a safe preservative for use in cosmetics. *Nanomedicine* **2010**, *6*, 570–574. [CrossRef] [PubMed]
6. De Moura, M.R.; Mattoso, L.H.; Zucolotto, V. Development of cellulose based bactericidal nanocomposites containing silver nanoparticles and their use as active food packaging. *J. Food Eng.* **2012**, *109*, 520–524. [CrossRef]
7. Echegoyen, Y.; Nerín, C. Nanoparticle release from nano-silver antimicrobial food containers. *Food Chem. Toxicol.* **2013**, *62*, 16–22. [CrossRef]
8. Noor, N.; Mutalik, S.; Younas, M.W.; Chan, C.Y.; Thakur, S.; Wang, F.; Yao, M.Z.; Mou, Q.; Leung, P.H. Durable antimicrobial behaviour from silver-graphene coated medical textile composites. *Polymers* **2019**, *11*, 2000. [CrossRef]
9. Lee, S.H.; Jun, B.-H. Silver nanoparticles: Synthesis and application for nanomedicine. *Int. J. Mol. Sci.* **2019**, *20*, 865. [CrossRef]
10. Murphy, M.; Ting, K.; Zhang, X.; Soo, C.; Zheng, Z. Current Development of silver nanoparticle preparation, investigation, and application in the field of medicine. *J. Nanomater.* **2015**, *2015*, 696918. [CrossRef]
11. Kumar, S.S.D.; Rajendran, N.K.; Houreld, N.N.; Abrahamse, H. Recent advances on silver nanoparticle and biopolymer-based biomaterials for wound healing applications. *Int. J. Biol. Macromol.* **2018**, *115*, 165–175. [CrossRef] [PubMed]
12. Braakhuis, H.M.; Gosens, I.; Krystek, P.; Boere, J.A.F.; Cassee, F.R.; Fokkens, P.H.B.; Post, J.A.; van Loveren, H.; Park, M.V.D.Z. Particle size dependent deposition and pulmonary inflammation after short-term inhalation of silver nanoparticles. *Part. Fibre Toxicol.* **2014**, *11*, 49. [CrossRef]
13. Chernousova, S.; Epple, M. Silver as antibacterial agent: Ion, nanoparticle, and metal. *Angew. Chem. Int. Ed.* **2013**, *52*, 1636–1653. [CrossRef]

14. Le Ouay, B.; Stellacci, F. Antibacterial activity of silver nanoparticles: A surface science insight. *Nano Today* **2015**, *10*, 339–354. [CrossRef]
15. Slavin, Y.N.; Asnis, J.; Häfeli, U.O.; Bach, H. Metal nanoparticles: Understanding the mechanisms behind antibacterial activity. *J. Nanobiotechnol.* **2017**, *15*, 65. [CrossRef] [PubMed]
16. Pareek, V.; Gupta, R.; Panwar, J. Do physico-chemical properties of silver nanoparticles decide their interaction with biological media and bactericidal action? A review. *Mater. Sci. Eng. C* **2018**, *90*, 739–749. [CrossRef]
17. Jung, W.K.; Koo, H.C.; Kim, K.W.; Shin, S.; Kim, S.H.; Park, Y.H. Antibacterial activity and mechanism of action of the silver ion in *Staphylococcus aureus* and *Escherichia coli*. *Appl. Environ. Microbiol.* **2008**, *74*, 2171–2178. [CrossRef]
18. Polivkova, M.; Strublova, V.; Hubacek, T.; Rimpelova, S.; Svorcik, V.; Siegel, J. Surface characterization and antibacterial response of silver nanowire arrays supported on laser-treated polyethylene naphthalate. *Mater. Sci. Eng. C* **2017**, *72*, 512–518. [CrossRef]
19. Yaqoob, A.A.; Umar, K.; Ibrahim, M.N.M. Silver nanoparticles: Various methods of synthesis, size affecting factors and their potential applications–A review. *Appl. Nanosci.* **2020**, *10*, 1369–1378. [CrossRef]
20. Iravani, S.; Korbekandi, H.; Mirmohammadi, S.V.; Zolfaghari, B. Synthesis of silver nanoparticles: Chemical, physical and biological methods. *Res. Pharm. Sci.* **2014**, *9*, 385–406.
21. Polivkova, M.; Hubacek, T.; Staszek, M.; Svorcik, V.; Siegel, J. Antimicrobial treatment of polymeric medical devices by silver nanomaterials and related technology. *Int. J. Mol. Sci.* **2017**, *18*, 419. [CrossRef]
22. Banne, S.V.; Patil, M.S.; Kulkarni, R.M.; Patil, S.J. Synthesis and characterization of silver nano particles for EDM applications. *Mater. Today* **2017**, *4*, 12054–12060. [CrossRef]
23. Roy, A.; Bulut, O.; Some, S.; Mandal, A.K.; Yilmaz, M.D. Green synthesis of silver nanoparticles: Biomolecule-nanoparticle organizations targeting antimicrobial activity. *RSC Adv.* **2019**, *9*, 2673–2702. [CrossRef]
24. Chintamani, R.B.; Salunkhe, K.S.; Chavan, M.J. Emerging use of green synthesis silver nanoparticle: An updated review. *Int. J. Pharm. Sci. Res.* **2018**, *9*, 4029–4055. [CrossRef]
25. Irshad, A.; Sarwar, N.; Sadia, H.; Malik, K.; Javed, I.; Irshad, A.; Afzal, M.; Abbas, M.; Rizvi, H. Comprehensive facts on dynamic antimicrobial properties of polysaccharides and biomolecules-silver nanoparticle conjugate. *Int. J. Biol. Macromol.* **2020**, *145*, 189–196. [CrossRef] [PubMed]
26. Lu, B.; Lu, F.; Zou, Y.; Liu, J.; Rong, B.; Li, Z.; Dai, F.; Wu, D.; Lan, G. In situ reduction of silver nanoparticles by chitosan-l-glutamic acid/hyaluronic acid: Enhancing antimicrobial and wound-healing activity. *Carbohydr. Polym.* **2017**, *173*, 556–565. [CrossRef]
27. Vosmanska, V.; Kolarova, K.; Rimpelova, S.; Kolska, Z.; Svorcik, V. Antibacterial wound dressing: Plasma treatment effect on chitosan impregnation and in situ synthesis of silver chloride on cellulose surface. *RSC Adv.* **2015**, *23*, 17690–17699. [CrossRef]
28. Hassabo, A.G.; Nada, A.A.; Ibrahim, H.M.; Abou-Zeid, N.Y. Impregnation of silver nanoparticles into polysaccharide substrates and their properties. *Carbohydr. Polym.* **2015**, *122*, 343–350. [CrossRef]
29. Ashraf, S.; Saif-ur-Rehman; Sher, F.; Khalid, Z.M.; Mehmood, M.; Hussain, I. Synthesis of cellulose–metal nanoparticle composites: Development and comparison of different protocols. *Cellulose* **2014**, *21*, 395–405. [CrossRef]
30. Vosmanska, V.; Kolarova, K.; Pislova, M.; Svorcik, V. Reaction parameters of in situ silver chloride precipitation on cellulose fibres. *Mater. Sci. Eng. C* **2019**, *95*, 134–142. [CrossRef]
31. Huang, H.H.; Ni, X.P.; Loy, G.L.; Chew, C.H.; Tan, K.L.; Loh, F.C.; Deng, J.F.; Xu, G.Q. Photochemical formation of silver nanoparticles in Poly(N-vinylpyrrolidone). *Langmuir* **1996**, *12*, 909–912. [CrossRef]
32. Lin, S.K.; Cheng, W.T. Fabrication and characterization of colloidal silver nanoparticle via photochemical synthesis. *Mater. Lett.* **2020**, *261*, 127077. [CrossRef]
33. Dong, C.; Zhang, X.; Cai, H. Green synthesis of monodisperse silver nanoparticles using hydroxy propyl methyl cellulose. *J. Alloys Compd.* **2014**, *583*, 267–271. [CrossRef]
34. Suwan, T.; Khongkhunthian, S.; Okonogi, S. Silver nanoparticles fabricated by reducing property of cellulose derivatives. *Drug Discov. Ther.* **2019**, *13*, 70–79. [CrossRef]
35. Walton, W.H. Feret's statistical diameter as a measure of particle size. *Nature* **1948**, *162*, 329–330. [CrossRef]

36. Kumbhar, A.S.; Kinnan, M.K.; Chumanov, G. Multipole plasmon resonances of submicron silver particles. *J. Am. Chem. Soc.* **2005**, *127*, 12444–12445. [CrossRef] [PubMed]
37. Joshi, S.C. Sol-Gel Behavior of Hydroxypropyl Methylcellulose (HPMC) in ionic media including drug release. *Materials* **2011**, *4*, 1861–1905. [CrossRef]

Publisher's Note: MDPI stays neutral with regard to jurisdictional claims in published maps and institutional affiliations.

coatings

MDPI

Article

Electrophoretic Deposition of Gentamicin-Loaded ZnHNTs-Chitosan on Titanium

Ahmed Humayun [1], Yangyang Luo [1] and David K. Mills [1,2,*]

1 Center for Biomedical Engineering and Rehabilitation Science, Louisiana Tech University, Ruston, LA 71272, USA; ah.humayun@gmail.com (A.H.); yangyang317luo@gmail.com (Y.L.)

2 School of Biological Sciences, Louisiana Tech University, Ruston, LA 71272, USA

* Correspondence: dkmills@latech.edu; Tel.: +1-318-257-2640

Received: 13 September 2020; Accepted: 30 September 2020; Published: 30 September 2020

check for updates

Abstract: There is a need for titanium (Ti), an antimicrobial implant coating that provides sustained protection against bacterial infection. Chitosan (CS) coatings, combined with halloysite nanotubes (HNTs), are an attractive solution due to the inherent biocompatibility of halloysite, its ability to provide sustained drug release, and the antimicrobial properties of CS. In this study, the electrodeposition (EPD) method was used to coat titanium foil with CS blended with zinc-coated HNTs (ZnHNTs) and pre-loaded with the antibiotic gentamicin. The CS-ZnHNTs-gentamycin sulfate (GS) coatings were characterized using scanning electron microscopy (SEM), energy-dispersive spectroscopy (EDS), X-ray powder diffraction (XRD), X-ray fluorescence (XRF), Fourier-transform infrared spectroscopy (FTIR), and UV-visible spectroscopy. The coatings were further examined for their ability to sustain GS release, resist bacterial colonization and growth, and prevent biofilm formation. The CS-ZnHNTs-GS coatings were cytocompatible, exhibited significant antimicrobial properties, and supported pre-osteoblast cell proliferation. Hydroxyapatite also formed on the coatings after immersion in simulated body fluid. While the focus in this study was on zinc-coated HNTs doped into CS, our design offers tunability, as different metals can be coated onto the HNT surface and different drugs or growth factors loaded into the HNT lumen. Our results, and the potential for customization, suggest that these coatings have potential in the construction of an array of infection-resistant implant coatings.

Keywords: antibiotics; drug loading; electrodeposition; halloysite nanotubes; zinc; metal nanoparticles; titanium implants

1. Introduction

Implant failure due to peri-implantitis and its prevention is a significant area of focus in implant-coating research. A high implant failure rate due to biofilm formation is a significant source of implant-related infections resulting in implant removal surgery, additional antibiotic treatment, higher medical costs, prolonged hospital stays, and increased mortality [1–4]. Biofilm formation is also a severe threat due to an increase in antimicrobial resistance, which poses a serious global threat to human communities. Several strategies have been developed to prevent biofilm formation, including the use of localized drug-releasing bioactive coatings, drug-doped nanocontainers [5,6], coating with metal nanoparticles [7,8], peptides [9], and quaternary ammonium salts [10]. For many decades, stainless steel was the material of choice for orthopedic implants. It has been replaced with titanium, primarily due to its strength, durability, corrosion resistance, and improved biocompatibility. Titanium, however, also has drawbacks such as low osseointegration and biofouling leading to post-surgical infections, resulting in a variety of complications. This unwanted patient clinical condition can be remedied by coating antibacterial materials on the implant surface [11].

Chitosan, a naturally occurring biopolymer, is extracted by the chitin's diacylation and comprises the insect exoskeleton [12,13]. It is biocompatible, biodegradable, non-toxic, and antibacterial, thus making it a useful component of many biomedical applications [14]. Halloysite is an aluminosilicate nanotube comprising rolled-up sheets of silica and alumina, resulting in an oppositely charged outer surface and an inner core [15,16]. It has shown significant potential as a drug carrier, as its hollow core can be loaded with various substances that can be released in a sustained manner [13,17]. Thus, biocompatible composites with a sustained drug release capability can be created by blending chitosan with drug-loaded halloysite nanotubes (HNTs).

Electrophoretic deposition (EPD) is a two-step coating method involving the movement of charged materials under an applied electrical field and the subsequent accumulation and deposition of the charged material on the oppositely charged electrodes [14]. Previous studies have applied EPD to fabricate surface coatings of titanium nanoparticles under different alcohol solutions [15], zinc-halloysite substituted hydroxyapatite [16] and a chitosan (CS)-HNT bioactive glass composite [6]. This study explores using EPD to deposit chitosan combined with zinc-coated halloysite (ZnHNT) loaded with gentamicin (CS-ZnHNTs-gentamycin sulfate (GS)) as an antibacterial titanium coating. The synergistic activity of gentamicin and metal ions in preventing bacterial growth and biofilm formation was examined as well as its cytocompatibility and impact on pre-osteoblast cell proliferation and functionality. CS-ZnHNTs-GS coatings were shown to be cytocompatible. The synergistic activity of gentamicin and metal ions exhibited significant antimicrobial properties. The prevention of biofilm formation was correlated with zinc ion release. Furthermore, these coatings also supported pre-osteoblast cell proliferation and functionality.

2. Materials and Methods

2.1. Materials

Staphylococcus aureus(*S. aureus*) ATCC® 6538™ 50 colony forming unit (CFU) (Manassas, VA, USA), chitosan (CS, *MW* = 200,000) with a deacetylation degree of 85%, HNTs, ethanol, acetic acid, acetone, phosphate-buffered saline (PBS), silicon carbide sandpaper, ninhydrin, gentamicin sulfate (GS), zinc sulphate heptahydrate ($ZnSO_4{\cdot}7H_2O$), titanium foil (Ti, 99.7% trace metals basis), o-ophthalaldehyde, β-mercaptoethanol, and isopropyl alcohol were purchased from Sigma-Aldrich (St. Louis, MO, USA). Platinum mesh electrodes, a VWR Accupower 500 electrophoresis power supply, and an ammeter (TP9605BT, TekPower, Seattle, WA, USA) were purchased from Amazon LLC.

2.2. Electrolytic Metallization of HNTs

ZnHNTs were prepared using a modified protocol based on an electrolytic method previously described by Mills et al. [17] (Figure 1). Briefly, an electrolysis setup was assembled consisting of two platinized titanium meshes held parallel at a 2-inch distance and connected to a DC power source (VWR Accupower 500 electrophoresis power supply). An ultrasonicated 100 mL aqueous solution of metal salts (2.5 mM) and 50 mg of HNTs were dispersed in glass beaker and subjected to 20 V at 80 °C with polarity reversal at every 5 min [17] under constant stirring to reduce electrophoretic buildup at the working electrode and thus increase the ion density in the solution; afterwards, the supernatant was decanted thrice, and the solution was centrifuged at 5000 rpm for 5 min with water to separate the ZnHNTs from non-adsorbed metal particles; they were dried at 30 °C.

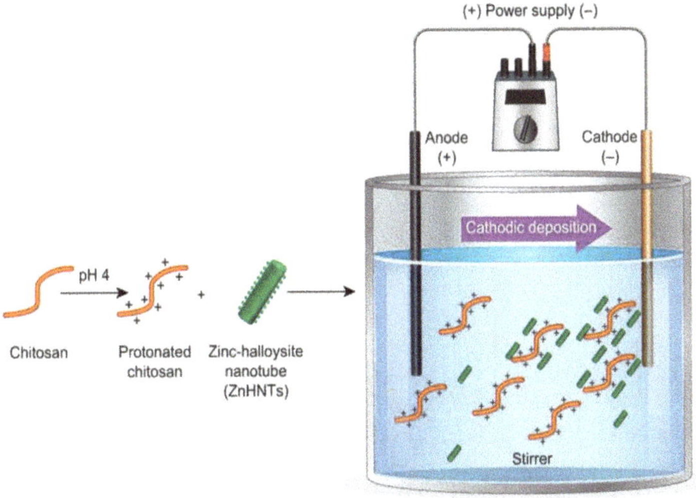

Figure 1. Graphical overview of the electrophoretic deposition process. An electrolysis setup consisting of two platinized titanium meshes held parallel at a 2-inch distance and connected to a DC power source. An ultrasonicated 100 mL aqueous solution of metal salts and HNTs were dispersed in a glass beaker and subjected to 20 V at 80 °C with polarity reversal at every 5 min [18] under constant stirring. Afterwards, the supernatant was decanted thrice, and the solution was combined with water to separate the mHNTs from non-adsorbed metal particles; they were dried at 30 °C.

2.3. Loading of HNTs with Gentamicin

A vacuum loading technique was employed to load gentamicin sulfate (GS) into the ZnHNT lumen. Briefly, 25 mg of GS was dispersed in 5 mL of PBS, mixed with 0.1 g of ZnHNTs, and placed under vacuum (28 pounds/square inch) overnight; the resulting product was decanted and washed. This process was repeated thrice to ensure the removal of free gentamicin.

2.4. Electrophoretic Deposition of the Coatings on Titanium

A 0.1 g solution of chitosan in 100 mL (3:1 distilled water/ethanol was prepared [6], and 1 mL of acetic acid was added in order to reduce the pH of the liquid, necessary for the protonation of chitosan, and 10 wt.% (0.01 g) GS-ZnHNTs were added. The solution was then magnetically stirred for 30 min [19].

Titanium foil was cut into 1 cm × 3 cm dimensions, grounded with silicon carbide paper, washed with acetone and distilled water successively in order to remove surface oil, and treated with 4 N NaOH for 10 min to increase the hydrophilicity of the surface; excess NaOH was wiped off, and the foil was rinsed in distilled water and dried at 30 °C. Electrophoretic deposition was performed using a platinum mesh electrode (5 cm × 7.5 cm) and titanium foil (1 cm × 3 cm) held 1 inch apart, connected to 40 V DC power supply (VWR Accupower 500 electrophoresis power supply), functioning as an anode and cathode with a desktop ammeter (TekPower TP9605BT) connected in series; they were immersed into the solution, and electrophoresis was carried out for 10 min under room temperature. The samples were air-dried, and the before and after weights were recorded (Figure 1).

2.5. Characterization of Titanium Coatings

A S-4800 field-emission scanning electron microscope (SEM, Hitachi, Tokyo, Japan) was used to examine the surface morphology of the coated HNTs and to visually confirm the presence of the metal coating appearing as clusters on the otherwise smooth outer surface of the HNTs. SEM-EDS was carried out with an EDAX energy dispersive X-ray analyzer linked to the Hitachi S-4800 SEM to evaluate the elemental composition and weight percentage (wt.%) deposition on the AgHNTs. EDS was operated at a working distance of 15 mm and an acceleration voltage of 15 kV; EDS spectra were analyzed using the EDAX Genesis software (6.43).

X-ray crystal diffraction analysis was performed on a Bruker D8 Venture diffractometer (Bruker, Karlsruhe, German) with Cu Kα1 radiation (λ = 1.5418 Å). The scan speed and step size used were 2 s and 0.02°, respectively; the diffraction patterns were recorded on a PW 1710 X-ray powder diffractometer (Philips, Amsterdam, Netherlands) over 2θ within 3° to 85°.

The samples were analyzed using a Thermo/ARLQuant'X energy dispersive X-ray fluorescence spectrometer. The X-ray tube was operated at 30 kV for 60 live seconds, using a 0.05 mm (thick) Cu primary beam filter in an air path for silver metal detection. The X-ray fluorescence (XRF) spectra were studied using the Wintrace 7.1™ software (Thermo Fisher Scientific, Waltham, MA, USA).

The infrared spectrum was recorded at a resolution of 4 s^{-1} with 16 scans using a Thermo Scientific NICOLET™ IR100 FT-IR Spectrometer (Thermo Fisher Scientific, Waltham, MA, USA). The Thermo Scientific OMNIC™ software was used to study the stretching bands.

Microplate absorbances were analyzed using a 800TS microplate reader (Biotek, Winooski, VT, USA) set at 630 nm absorbance.

2.6. Gentamicin Release Analysis

The drug release quantity was determined over a six-day period using a method previously described [20]. Briefly, Ti samples were placed in 1 mL of PBS and placed in a shaker, and 1 mL of sample was collected at regular time intervals and 1 mL PBS was replenished each time; 1 mL of collected PBS was mixed with o-phthaldialdehyde (OPTA) reagent (1 mL) and isopropyl alcohol (1 mL) and allowed to react for 30 min at room temperature, and the absorbance was measured at 335 nm. OPTA regent was prepared by mixing 2.5 g of o-ophthalaldehyde, 3 mL of β-mercaptoethanol, 62. mL of isopropanol, and 560 mL of 0.04 sodium borate. PBS was used as the blank, and the measured quantities of drugs were used to draw standard graphs [21].

2.7. Antibacterial Analysis

S. aureus was used in this study; it was maintained in tryptic soy agar. For testing, the bacterial strain was cultured in nutrient broth and plated on Müller–Hinton agar plates at 37 °C overnight, after which a single colony was picked up using a sterile toothpick, suspended in saline solution, and diluted to 0.5 McFarland standards (1.5 × 10^8 CFU/mL), 20 µL of which was spread over Müller–Hinton agar plates on which the test materials were placed and incubated for 12–18 h at 37 °C, and the obtained zones of inhibition were analyzed using the ImageJ software.

The antibacterial potential was also evaluated for *S. aureus* using a microdilution broth assay. Briefly, 1 cm × 1 cm samples were cut and immersed in 24-well plates containing 3 mL/well of Müller–Hinton broth inoculated with 20 µL of 0.5 McFarland standard *S. aureus;* the plates were put on a shaker; after 4 h, the samples were removed, and the absorbance of 100 µL of the solutions was recorded after 12 h.

A bacterial adherence assay was performed by immersing samples into 1–3 mL/well of Müller–Hinton broth (depending on the sample dimension) inoculated with 20 µL of 0.5 McFarland standard *S. aureus*, the plates were put on a shaker, and the samples were removed after 1 h; the absorbance of 100 µL of the solutions was recorded after 12 h [22].

A biofilm assay was performed using the crystal violet assay. Briefly, samples were incubated in 2 mL of nutrient broth in 48-well plates inoculated with 1 OD *S. aureus* for 2 days at 37 °C, at the end of which the plate was emptied by inverting with gentle tapping in order to remove lightly attached planktonic bacteria. The remaining bacterial films were stained with aqueous crystal violet (0.1% *w*/*v*) for 10 min, which was similarly removed by inversion and tapping to ensure the optimal removal of unattached bacteria. Acetic acid (30%) was then added to each well to solubilize the stains, and the absorbance at 630 nm was recorded [23].

2.8. In Vitro Bioactivity Tests

In vitro bioactivity was assessed by using 5× simulated body fluid (SBF) prepared using a modified protocol [24] based on the method of Kokubo et al. [25]. The samples were immersed in 3 mL of 5 × SBF at 37 °C for 1 week; after immersion, the samples were taken out and dried at room temperature. SEM-EDS was performed on the samples to detect the formation of hydroxyapatite (HA).

2.9. Cell Culture

1 cm × 1 cm samples were sterilized in a UV chamber for 60 min. Cell viability was studied by co-culturing pre-osteoblast cells (MC3T3) with mHNTs. The cells were obtained from the American Type Culture Collection (ATCC, Manassas, VA, USA). Cryovials were thawed and allowed to equilibrate in a water bath within a humidified CO_2 incubator at 37 °C. The cells thus obtained were cultured in T25 flasks in alpha MEM growth media (Hyclone, GE Life Sciences, Marlborough, MA, USA) containing 10% FBS and 1% penicillin (complete medium) through to passage four and then frozen down and maintained in a liquid nitrogen Dewar until use. Trypsin-EDTA (0.25% trypsin, 1 mM EDTA) was used to detach the cells from the culture flasks.

For each experiment (n = 3), pre-osteoblasts were thawed and prepared as described above and used when they subsequently achieved sub-confluency. For every cell study, pre-osteoblast cells were seeded in 48-well plates at a concentration of 1×10^5 cells/ well. After 24 h of incubation, 25 μg/mL metalized HNTs (mHNTs) were added into each well for another 24 h of incubation. Then, the cells were used to assess the potential cytotoxicity (Live/Dead assay) and cell proliferation (MTS assay).

2.10. Live/Dead Cytotoxicity Tests and Cell Proliferation Studies

Ti implants with different coatings were placed into 48-well plates. Each coating group had 3 samples. Pre-osteoblast cells were seeded into each well at a concentration of 1×10^5 cells/well. Cells cultured in wells without Ti implants were used as controls. Cells were cultured at 37 °C for 24 h. Cell viability was assessed using the LIVE/DEAD® Viability/Cytotoxicity Kit (ThermoFisher, Waltham, MA, USA); it contains the polyanionic dye calcein, which is retained within living cells and produces an intense uniform green fluorescence (excitation/emission ~495 nm/~515 nm), and EthD-1, which enters cells with damaged membranes and produces a bright red fluorescence in dead cells (excitation/emission ~495 nm/~635 nm).

For viability studies, the staining solution was prepared by mixing 5 μL of 4 mM Calcein AM Solution and 20 μL of 2 mM EthD-III Solution with 10 mL of DPBS. Cells were washed twice with DBSS, and then, 50 μL of staining solution was added to each well. Then, the cells were incubated at room temperature for 30 min in darkness. Each cytotoxicity experiment was repeated three times. Images were captured using an Olympus BX51 fluorescence microscope (Olympus Corporation, Tokyo, Japan) equipped with an Olympus DP11 digital camera system. ImageJ [26] was used for counting fluorescent cells from thresholded 8-bit images at a cell size limit of 50–200 pixels^2.

Cell proliferation was assessed with the MTS colorimetric assay, which is based on the reduction of the MTS tetrazolium compound to a colored formazan product soluble in culture media. Cells were cultured as above for 24 h, and then, 40 μL of MTS reagent was added to each well and incubated at 37 °C for 2 h. The absorbance value at 490 nm was recorded; the background absorbance was determined from the medium wells, which only contained cell culture medium, and subtracted.

2.11. Statistical Analysis

All experiments were performed in triplicate, and the results are reported as mean ± standard deviation. One-way analysis of variance (ANOVA) with $p < 0.05$ as the significance level was utilized for statistical analysis. Statistical analysis was performed using the Microsoft Excel Analysis ToolPak plugin and Origin 9.6. Linear regression was used to construct and correlate standard curves.

3. Results and Discussion

GS adsorbed to the outer surface of the ZnHNTs due to its Si–O–Si negatively charged bonds, whereas the application of a vacuum creates negative pressure inside the ZnHNT lumen and facilitates drug loading [27]. Chitosan is necessary for the EPD of ZnHNTs-GS, as it not only stabilized the suspension but also provided the required net positive charge for deposition on the negatively charged cathode [5]. Chitosan dissolves by protonation through acetic acid addition, resulting in pH reduction, increased conductivity, and adsorption onto the ZnHNTs [19].

$$CH_3COOH + CS - NH_2 \leftrightarrow CH_3COO^- + CS - NH_3^+ \quad (1)$$

At the cathode, the electrolysis of water occurs, generating the hydroxyl ion (OH^-), which deprotonates the CS-NH_3^+, resulting in CS-NH_2 [5,28,29]. Several recent studies have examined the use of GS as an HNT coating, with several ingenious methods developed for their use in drug delivery [28,29]. HNTs wrapped by chitosan layers appear to be a very effective means for drug delivery.

Samples examined using laser confocal microscopy revealed a thin layer of the CS coated on the surface of the Ti foil (Figure 2).

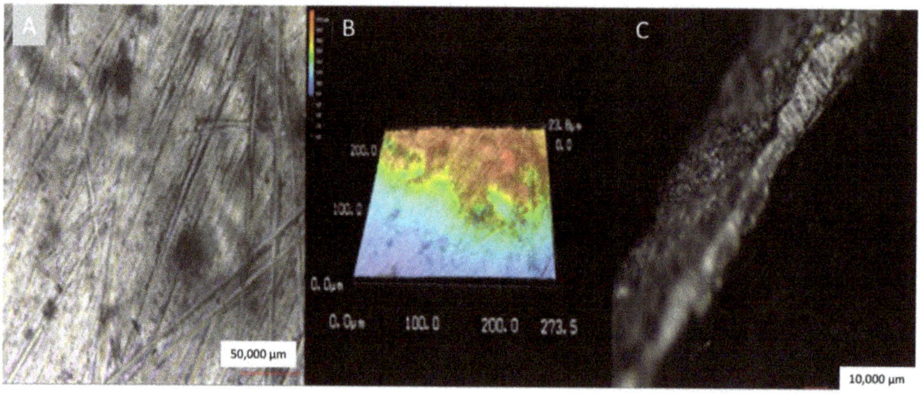

Figure 2. (**A**,**C**) Laser confocal image of chitosan (CS)-ZnHNTs-gentamycin sulfate (GS) coating; (**B**) 3D topography of the coated Ti foil.

As shown in the SEM micrograph in Figure 3, a uniform layer of chitosan was observed on all samples. A thin layer of chitosan with circular areas with no deposition was electrodeposited on titanium; the smooth morphology of the chitosan (Figure 3A) was disrupted with the addition of mHNTs that appeared uniformly distributed across the chitosan-coated Ti surface (Figure 3B). The ZnHNTs exhibited a circular morphology on the surface, indicative of the presence of HNTs with varyingly sized ZnHNTs visible across the Ti surface.

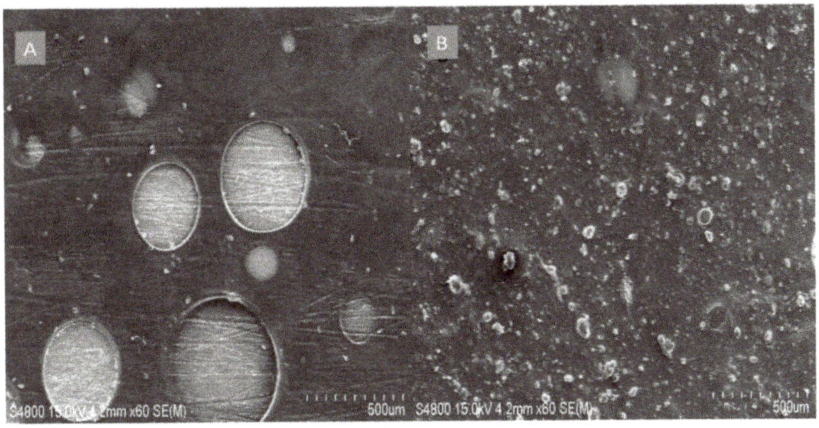

Figure 3. SEM micrograph of (**A**) chitosan and (**B**) CS-ZnHNTs-GS deposits.

In Figure 4, the elemental mapping results confirmed the presence of uniformly interspersed ZnHNTs distributed throughout the coatings. Furthermore, in the EDS spectra, strong silicon, aluminum, and zinc peaks were observed confirming their presence.

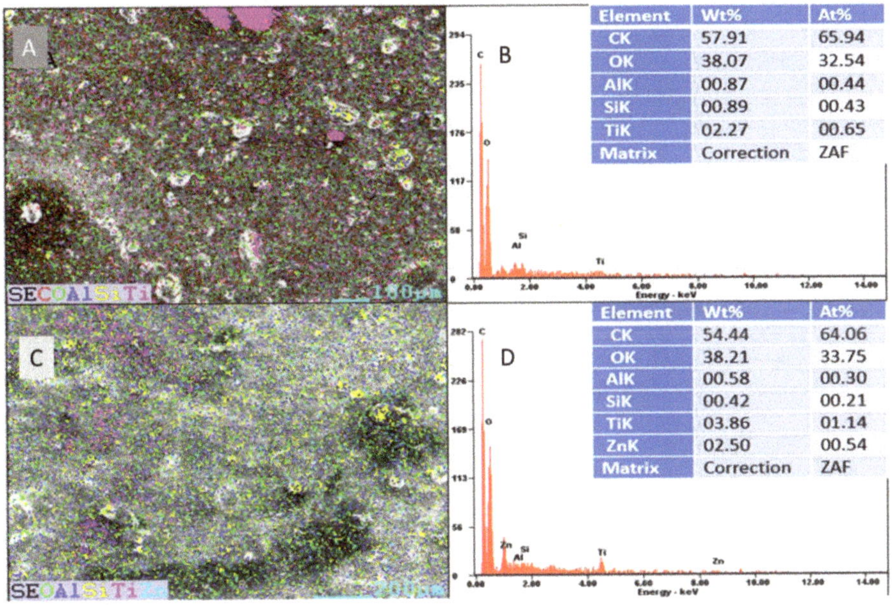

Figure 4. EDS mapping of (**A**,**B**) CS-HNTs and (**C**,**D**) CS-ZnHNTs; a more uniform surface with less titanium base metal visibility was obtained, which could be due to the increased conductance of the electrophoretic deposition upon the addition of ZnHNTs.

FTIR was used for chemical analysis and the identification of the characteristic bands for chitosan at 3400 (O–H stretch) and 2940 cm^{-1} (C–H stretch), respectively, and peaks for the secondary amide group bending were observed at 1644 cm^{-1}. Upon the addition of ZnHNTs, new peaks were observed at 1031 and 1090 cm^{-1}, respectively, which correspond to the in-plane Si–O stretching of HNTs, with new peaks at 1278 and 1401 cm^{-1}, respectively, that can be ascribed to the C–N stretching of gentamicin [30] (Figure 5).

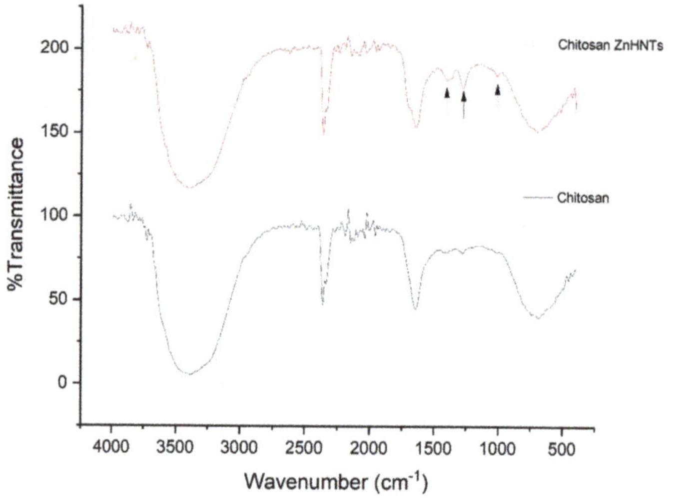

Figure 5. FTIR spectra of the coatings. Arrow depicts the new peaks observed upon the addition of ZnHNTs-GS at 1028 (in-plane Si–O stretching of HNTs), 1278 and 1401 cm^{-1} corresponding to the C–N stretching of gentamicin.

In the XRD spectra of the coatings, a broad peak at 20° was observed, which is attributable to the crystalline structure of chitosan (denoted by *), and peaks at 12, 20, and 25° were observed for the (001), (020), and (002) planes of HNTs. CS-ZnHNTs displayed the peaks corresponding to ZnO at 31°, 34°, 36°, 47°, 56°, 63°, and 68° corresponding to the (100), (002), (101), (102), (110), (103), and (112) planes, respectively (Figure 6) [31,32]. Figure 7 shows the current density for the EPD at 40 V in an ethanolic solution, and it remained almost constant during the process. Furthermore, the addition of acetic acid helped in chitosan protonation and increased with the current. The addition of ZnHNTs resulted in an elevated amperage.

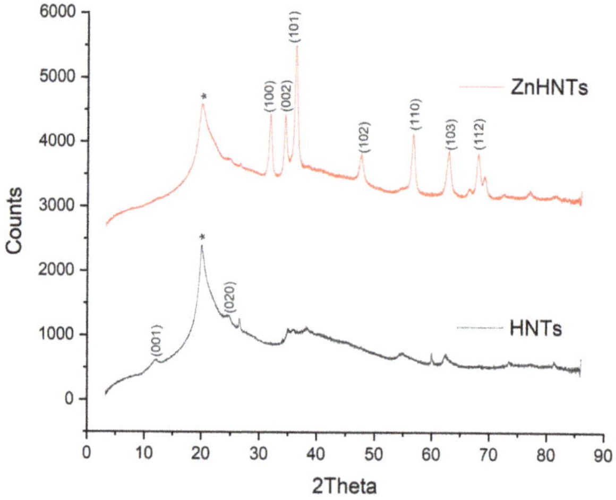

Figure 6. XRD profile of CS-HNTs and CS-ZnHNTs.

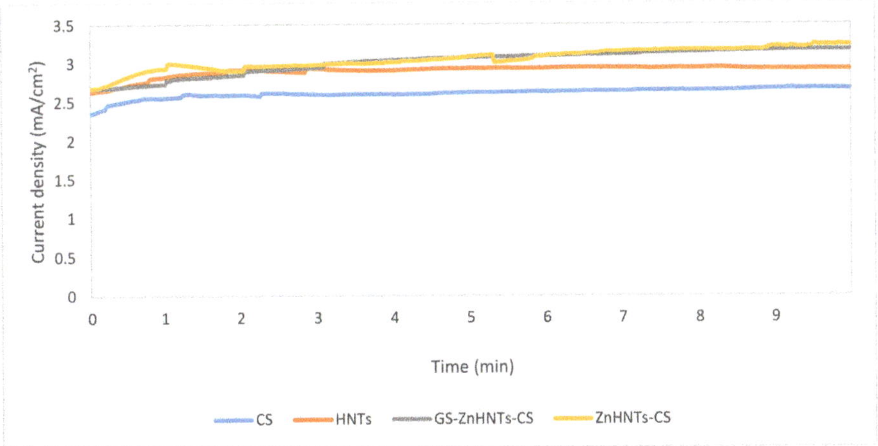

Figure 7. Current density during electrodeposition (EPD) at 40 V. The addition of ZnHNTs in the chitosan leads to increased current density, explainable based on increased conductivity due to ionic species leaching from ZnHNTs.

The bacterial infection of implants is the leading cause of implant failure [33]. Bacterial adhesion, growth, and biofilm formation after implantation can induce severe complications for the patient [34]. Different approaches for producing antimicrobial medical devices have been studied. Surface functionalization, with different components, including antibiotics, antimicrobial dyes, quants, and other antibacterial bioactives, has been a significant research focus [30]. Metal nanoparticles exhibit robust antimicrobial action, especially copper, silver, and zinc oxide. Several recent studies have used coating solutions containing CS and metal nanoparticles. Copper [35], silver [36], titanium dioxide [37], and zinc [38] were the most commonly used metal nanoparticles.

The present study had a very similar motive, with a goal of developing a biomaterial surface coating possessing antimicrobial capability and, in this case, studying the combined effect on a known antimicrobial nanoparticle (zinc) and gentamycin-doped HNTs. The antibacterial potential of the coatings was evaluated for Gram-positive >*S. aureus* bacteria using the agar diffusion method. CS and HNTs alone did not induce an antibacterial response (Figure 8A,B), while ZnHNTs-CS produced a large inhibitory effect (Figure 8C). By contrast, CS-ZnHNTs-GS exhibited significant inhibition zones of 3.11 ± 0.79 cm^2/unit area of the sample (Figure 8D), and a similar pattern was observed in the antimicrobial broth testing (Figure 8E). A similar bacterial adhesion assay showed similar antibacterial activity but a diminished trend as compared to that shown by broth testing (Figure 9B). In addition, fewer adherent planktonic bacteria were found (Figure 9B).

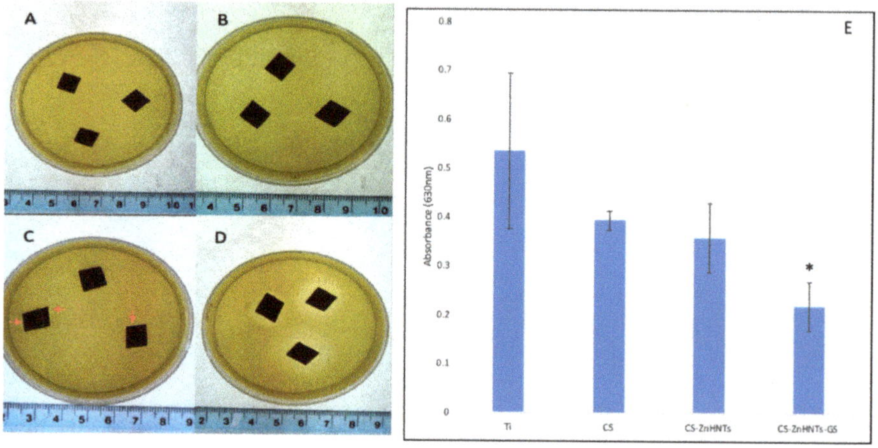

Figure 8. Disc diffusion assay for *S. aureus*. (**A**) CS, (**B**) HNTs-CS, (**C**) ZnHNTs-CS, and (**D**) GS-ZnHNTs-CS. (**E**) Antimicrobial broth testing. GS-ZnHNTs-CS exhibited the highest bacteriostatic effect. Bars = mean ± standard deviation. * = significant difference, $p < 0.05$ ($n = 3$).

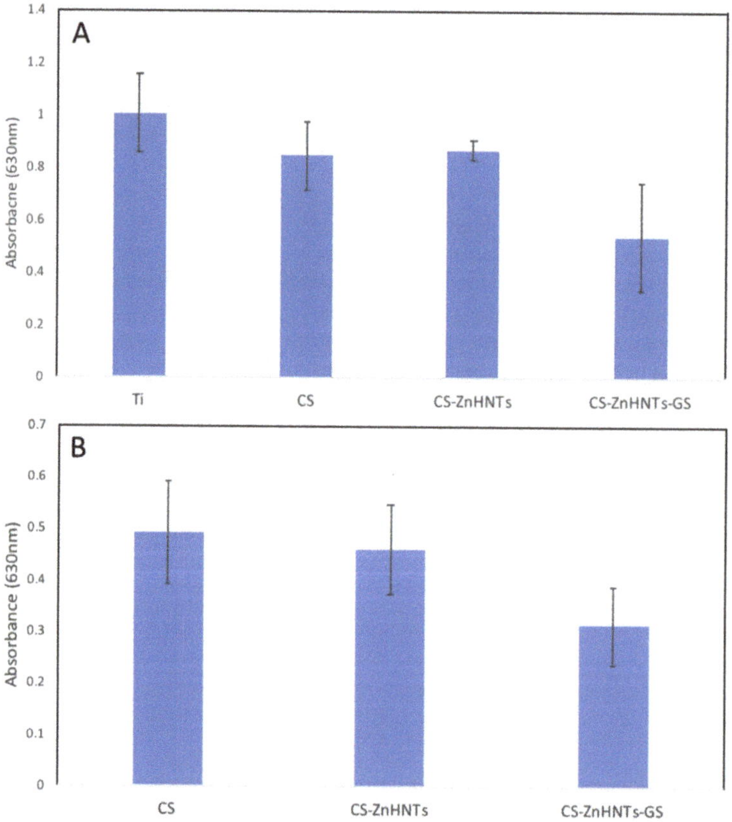

Figure 9. (**A**) Bacterial adherence assay using *S. aureus* applied to the surface of the samples. (**B**) Biofilm assay using crystal violet.

Crystal violet is a basic dye that binds to the peptidoglycan layers of Gram-positive and negative bacteria [39]. The samples on staining with crystal violet displayed less biofilm formation on the implant surface on the incorporation of ZnHNTs-GS (Figure 10); this may be attributed to zinc ions leaching from the coatings (Figure 11) and the release of gentamicin from the HNT lumen (Figure 12). Ti samples were placed in 3 mL PBS, placed in an orbital shaker and 3 mL of sample was analyzed at regular intervals of time (1, 24, 48, 72 h), the PBS sample was restored to the original container after XRF analysis.

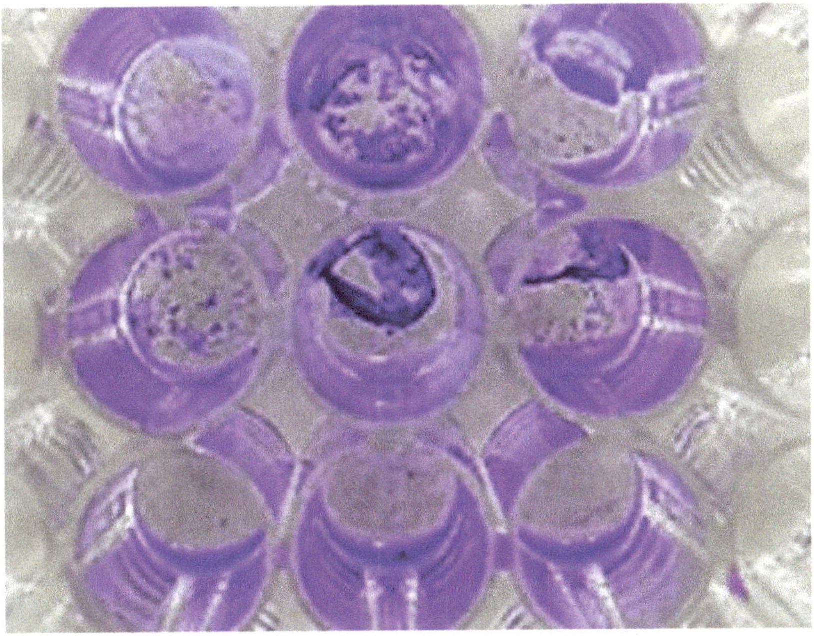

Figure 10. Crystal violet-stained biofilms in CS, CS-ZnHNTs, and CS-ZnHNTs-GS wells (top–bottom, respectively).

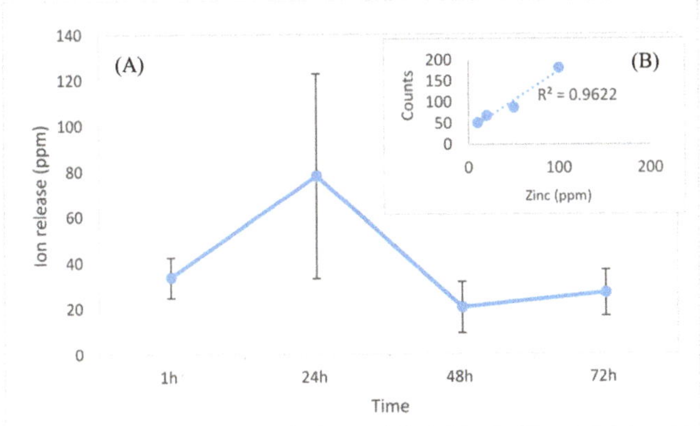

Figure 11. (**A**) Zinc ion release from the CS-ZnHNTs-GS coatings; highest ion release concentration was observed at 24 h (n = 5), (**B**—zinc ion standard X-ray fluorescence (XRF) curve).

Additionally, the total GS content on the surface of the implant coatings was estimated by dipping the samples into a ninhydrin reagent (2.12 ± 0.81 mg/mL was obtained, n = 5). The data show that the gentamicin-loaded ZnHNTs-chitosan coatings possesses antibacterial activity, which can be attributed mainly to the gentamicin antibiotic. Additionally, the release of zinc ions, as well as the positively charged chitosan surface, leads to interaction with the negatively charged bacterial membranes, further enhancing the antibacterial effect.

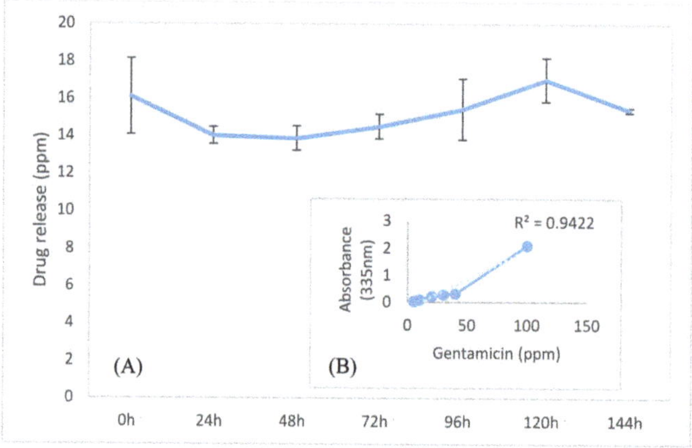

Figure 12. (**A**) Gentamicin release from CS-ZnHNTs-GS coatings (ppm). A gradual burst of drug release was observed, which could be due to drug adsorption on the mHNTs' outer surface and chitosan coating. (**B**—gentamicin standard XRF curve.) (n = 5.)

Cytocompatibility was assessed using a Live/Dead assay. As shown in Figure 13, there were few dead cells detected on the substrates containing ZnHNTs (Figure 13(3b,4b)). The total numbers of live and dead cells were counted using the ImageJ software, and the total cell viability was plotted, as shown in Figure 14. An ANOVA analysis did not show significant differences between each coating type and the control group; however, the CS-ZnHNTs and CS-ZnHNTs-GS groups exhibited slightly higher cell viability. One possible explanation for this observation is the Zn^+ ion release. Zinc is an essential element in human health, has been found as a component in over 100 enzymes, and also plays a crucial role in cell proliferation and DNA synthesis [40]. Additionally, zinc is an antioxidant and helps to prevent free radical-induced injury [41]. In this study, as zinc was released from the HNT surface, it was taken up by surrounding cells, contributing to enhanced cell viability.

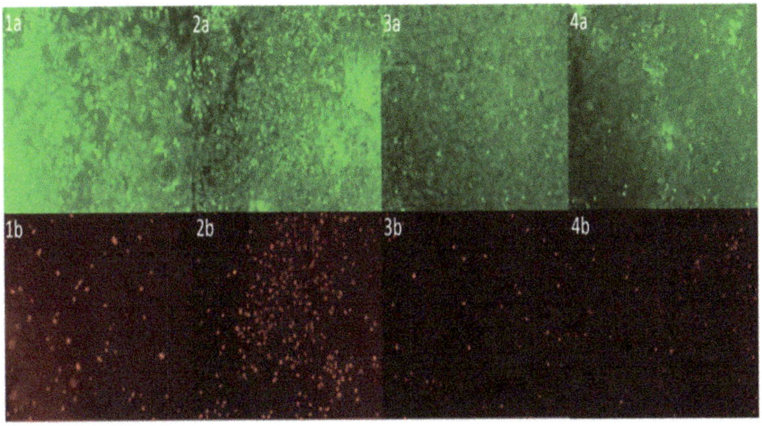

Figure 13. Live/Dead assay; pre-osteoblast cells were cultured in the presence of Ti foil scaffolds: (1**a**,**b**) CS, (2**a**,**b**) HNTs-CS, (3**a**,**b**) ZnHNTs-CS, and (4**a**,**b**) GS-ZnHNTs-CS. Live cells (green) and dead cells (red). ZnHNTs led to increased cellular viability.

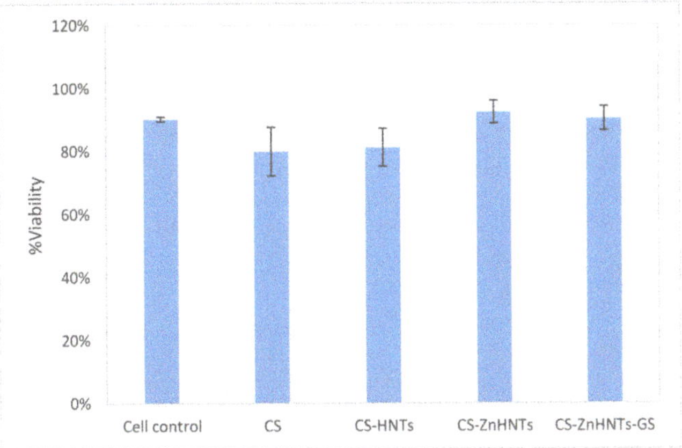

Figure 14. Cell viability assay for Ti foil scaffolds; the CS-ZnHNTs coating had the highest cell viability (92%).

Cell proliferation in response to CS-ZnHNTs and CS-ZnHNTs-GS was analyzed using the MTS assay. As shown in Figure 15, when compared to the control group, the cells cultured on all the zinc/HNT-coated Ti implants showed a reduction in cell proliferation, and after 24 h of incubation, the cells grown on Ti implants exhibited reduced proliferation. This was an unexpected observation. One explanation may be deficient surface cell adhesion properties in the zinc-coated Ti groups that led to reduced cell attachment on the coated Ti surface after plating. Cell culture plates are treated to promote cell attachment, and their surface contains hydroxyl and carboxyl groups, which promotes cell attachment. In this study, the coated Ti implants did not include any specific cell adhesion peptides or ligands to promote cell attachment and the CS was unmodified, and this may have led to a reduced number of cells adhering to the Ti surfaces during initial cell seeding. Many studies have modified CS through plasma treatment to encourage cell attachment [42]. The surface treatment of CS is an effective way of improving the biocompatibility of chitosan membranes for their use in biomedical applications.

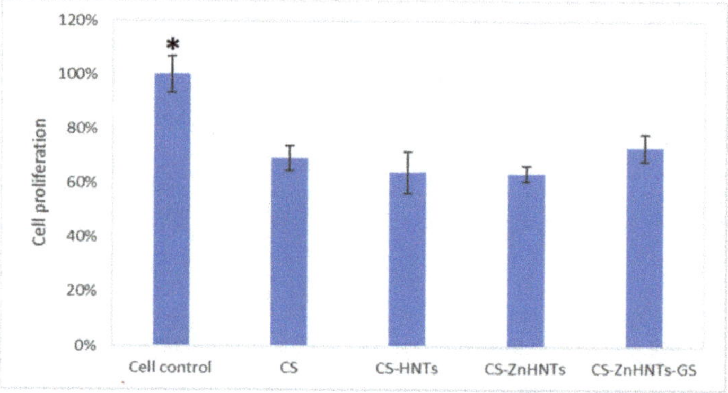

Figure 15. Proliferation assay for the Ti foil scaffolds after 24 h of incubation. * = statistically significant at $p < 0.05$.

Cell proliferation on Ti implant coatings improved after 24 h, and after an increased incubation time, cell proliferation was similar to that in the control group, demonstrating that all the coatings are cytocompatible (Figure 15); the presence of Zn may promote cell proliferation. It has been reported that the addition of Zn led to increased bone marrow stem cell proliferation [43]. Another study showed that Zn increased antimicrobial and osteogenic activity [44], and similar results have been found when Zn was incorporated with magnesium ions [45].

Cumulatively, the data show that the CS-ZnHNTs-GS coatings possess significant antibacterial activity, which can be attributed largely to the release of gentamicin. Additionally, the release of zinc ions, as well as the positively charged chitosan surface, might have led to increased interactions with negatively charged bacterial membranes and resulted in the supplementary enhancement of the antibacterial effect. Increased cellular viability and proliferation were also observed with exposure to ZnHNTs. Hydroxyapatite was also observed to have formed on the samples after immersion in SBF (Figure 16). As an essential mineral required for normal skeletal growth and bone maintenance, zinc can positively affect chondrocyte and osteoblast functions, while inhibiting osteoclast activity [46,47]. Hydroxyapatite formation is an indication of the potential of CS-ZnHNT coatings in promoting bone formation leading to osseointegration [47–49]. However, further investigation of its potential to induce osteoblast differentiation leading to bone tissue formation and mineralization will be required to validate its potential.

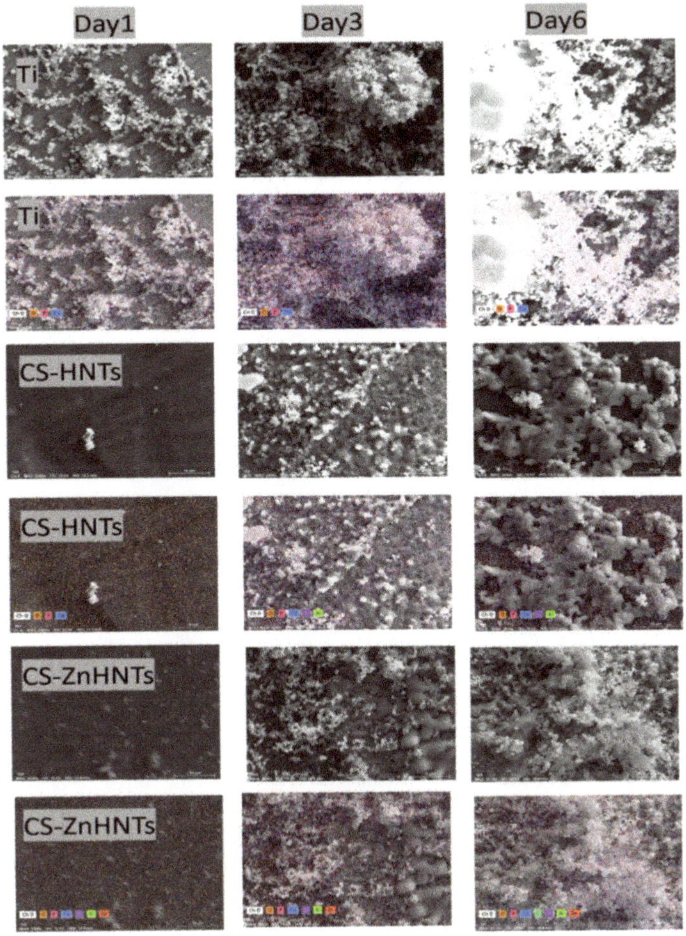

Figure 16. SEM and SEM-EDS map of the samples soaked in simulated body fluid (SBF) for 1, 3, and 6 days.

4. Conclusions

Electrodeposition was used to deposit gentamicin-loaded ZnHNTs-chitosan coatings on titanium foil. Morphological surface characterization using SEM-EDS and EDS mapping confirmed the presence of tubular ZnHNTs structures embedded in the coatings, and FTIR analysis confirmed the presence of GS-ZnHNTs-CS in the coatings. Bacterial growth inhibition assays confirmed the successful development of an antibacterial coating. The coating showed cytocompatibility and enhanced cell proliferation in pre-osteoblast cells.

Author Contributions: The authors all contributed to the writing of the manuscript. A.H. and Y.L. conducted the experiments under the direction of D.K.M. All authors reviewed and analyzed the data. All authors have read and agreed to the published version of the manuscript.

Funding: Funding for this study was provided by a grant (to Dr. Mills) from the Louisiana Biomedical Research Network (through an Institutional Development Award (IDeA) from the National Institute of General Medical Sciences of the National Institutes of Health under grant number P20 GM103424-17). The authors also wish to acknowledge the support of the College of Applied and Natural Sciences (Louisiana Tech University) through the Matching Grant Program.

Acknowledgments: Davis Bailey and Sven Eklund for training.

Conflicts of Interest: The authors declare no conflict of interest.

References

1. Mattioli-Belmonte, M.; Cometa, S.; Ferretti, C.; Iatta, R.; Trapani, A.; Ceci, E.; Falconi, M.; De Giglio, E. Characterization and cytocompatibility of an antibiotic/chitosan/cyclodextrins nanocoating on titanium implants. *Carbohydr. Polym.* **2014**, *110*, 173–182. [CrossRef] [PubMed]
2. Ordikhani, F.; Dehghani, M.; Simchi, A. Antibiotic-loaded chitosan–Laponite films for local drug delivery by titanium implants: Cell proliferation and drug release studies. *J. Mater. Sci. Mater. Electron.* **2015**, *26*, 269. [CrossRef] [PubMed]
3. Norowski, P.A.; Courtney, H.S.; Babu, J.; Haggard, W.O.; Bumgardner, J.D. Chitosan coatings deliver antimicrobials from titanium implants: A preliminary study. *Implant. Dent.* **2011**, *20*, 56–67. [CrossRef] [PubMed]
4. Swanson, T.E.; Cheng, X.; Friedrich, C. Development of chitosan-vancomycin antimicrobial coatings on titanium implants. *J. Biomed. Mater. Res. Part A* **2011**, *97*, 167–176. [CrossRef] [PubMed]
5. Farrokhi-Rad, M.; Fateh, A.; Shahrabi, T. Electrophoretic deposition of vancomycin loaded halloysite nanotubes-chitosan nanocomposite coatings. *Surf. Coat. Technol.* **2018**, *349*, 144–156. [CrossRef]
6. Radda'A, N.S.; Goldmann, W.H.; Detsch, R.; Roether, J.A.; Cordero-Arias, L.; Virtanen, S.; Moskalewicz, T.; Boccaccini, A.R. Electrophoretic deposition of tetracycline hydrochloride loaded halloysite nanotubes chitosan/bioactive glass composite coatings for orthopedic implants. *Surf. Coat. Technol.* **2017**, *327*, 146–157. [CrossRef]
7. Besinis, A.; Hadi, S.D.; Le, H.R.; Tredwin, C.; Handy, R.D. Antibacterial activity and biofilm inhibition by surface modified titanium alloy medical implants following application of silver, titanium dioxide and hydroxyapatite nanocoatings. *Nanotoxicology* **2017**, *11*, 327–338. [CrossRef]
8. Ewald, A.; Glückermann, S.K.; Thull, D.-I.R.; Gbureck, U. Antimicrobial titanium/silver PVD coatings on titanium. *Biomed. Eng. Online* **2006**, *5*, 22. [CrossRef]
9. Kazemzadeh-Narbat, M.; Lai, B.F.; Ding, C.; Kizhakkedathu, J.N.; Hancock, R.E.W.; Wang, R. Multilayered coating on titanium for controlled release of antimicrobial peptides for the prevention of implant-associated infections. *Biomaterials* **2013**, *34*, 5969–5977. [CrossRef]
10. Peng, Z.-X.; Tu, B.; Shen, Y.; Du, L.; Wang, L.; Guo, S.-R.; Tang, T.-T. Quaternized chitosan inhibits icaA transcription and biofilm formation by Staphylococcus on a titanium surface. *Antimicrob. Agents Chemother.* **2011**, *55*, 860–866. [CrossRef]
11. Prasad, K.; Bazaka, O.; Chua, M.; Rochford, M.; Fedrick, L.; Spoor, J.; Symes, R.; Tieppo, M.; Collins, C.; Cao, A.; et al. Metallic biomaterials: current challenges and opportunities. *Materials* **2017**, *10*, 884. [CrossRef] [PubMed]
12. Anitha, A.; Sowmya, S.; Jayakumar, R.; Deepthi, S.; Chennazhi, K.; Ehrlich, H.; Tsurkan, M.V.; Jayakumar, R. Chitin and chitosan in selected biomedical applications. *Prog. Polym. Sci.* **2014**, *39*, 1644–1667. [CrossRef]
13. Massaro, M.; Lazzara, G.; Milioto, S.; Noto, R.; Riela, S. Covalently modified halloysite clay nanotubes: Synthesis, properties, biological and medical applications. *J. Mater. Chem. B* **2017**, *5*, 2867–2882. [CrossRef] [PubMed]
14. Augello, C.; Liu, H. Surface modification of magnesium by functional polymer coatings for neural applications. *Surf. Modif. Magnes. Alloys Biomed. Appl.* **2015**, 335–353. [CrossRef]
15. Farrokhi-Rad, M.; Ghorbani, M. Electrophoretic deposition of titania nanoparticles in different alcohols: Kinetics of deposition. *J. Am. Ceram. Soc.* **2011**, *94*, 2354–2361. [CrossRef]
16. Chozhanathmisra, M.; Ramya, S.; Kavitha, L.; Gopi, D. Development of zinc-halloysite nanotube/minerals substituted hydroxyapatite bilayer coatings on titanium alloy for orthopedic applications. *Colloids Surf. A Physicochem. Eng. Asp.* **2016**, *511*, 357–365. [CrossRef]
17. Mills, D.; Boyer, C. Method for Metalizing Nanotubes through Electrolysis. U.S. Patent 9,981,074, 29 May 2018.
18. Khaydarov, R.A.; Khaydarov, R.R.; Gapurova, O.; Estrin, Y.; Scheper, T. Electrochemical method for the synthesis of silver nanoparticles. *J. Nanopart. Res.* **2008**, *11*, 1193–1200. [CrossRef]
19. Mahmoodi, S.; Sorkhi, L.; Farrokhi-Rad, M.; Shahrabi, T. Electrophoretic deposition of hydroxyapatite–chitosan nanocomposite coatings in different alcohols. *Surf. Coat. Technol.* **2013**, *216*, 106–114. [CrossRef]

20. Ismail, A.F.H.; Mohamed, F.; Rosli, L.M.M.; Shafri, M.A.M.; Haris, M.S.; Adina, A.B. Spectrophotometric determination of gentamicin loaded PLGA microparticles and method validation via ninhydrin-gentamicin complex as a rapid quantification approach. *J. Appl. Pharm. Sci.* **2016**, *6*, 7–14. [CrossRef]
21. Weisman, J.A.; Jammalamadaka, U.; Tappa, K.; Mills, D.K. Doped halloysite nanotubes for use in the 3D printing of medical devices. *Bioengineering* **2017**, *4*, 96. [CrossRef]
22. Lima, E.M.C.X.; Koo, H.; Smith, A.M.V.; Rosalen, P.L.; Cury, A.A.D.B. Adsorption of salivary and serum proteins, and bacterial adherence on titanium and zirconia ceramic surfaces. *Clin. Oral Implant Res.* **2008**, *19*, 780–785. [CrossRef] [PubMed]
23. O'Toole, G.A. Microtiter dish biofilm formation assay. *J. Vis. Exp.* **2011**, e2437. [CrossRef] [PubMed]
24. Barrere, F.; Van Blitterswijk, C.; De Groot, K.; Layrolle, P.; Van Blitterswijk, C.A. Nucleation of biomimetic Ca–P coatings on Ti6Al4V from a SBF×5 solution: Influence of magnesium. *Biomaterials* **2002**, *23*, 2211–2220. [CrossRef]
25. Kokubo, T.; Takadama, H. How useful is SBF in predicting in vivo bone bioactivity? *Biomaterials* **2006**, *27*, 2907–2915. [CrossRef] [PubMed]
26. Schneider, C.A.; Rasband, W.S.; Eliceiri, K.W. NIH Image to ImageJ: 25 years of image analysis. *Nat. Methods* **2012**, *9*, 671–675. [CrossRef] [PubMed]
27. Patel, S.; Jammalamadaka, U.; Sun, L.; Tappa, K.; Mills, D.K. Sustained telease of antibacterial agents from doped halloysite nanotubes. *Bioengineering* **2015**, *3*, 1. [CrossRef] [PubMed]
28. Zhitomarsky, I.; Gal-Or, L. Electrophoretic deposition of hydroxyapatite. *J. Mater. Sci. Mater. Med.* **1997**, *8*, 213–219. [CrossRef]
29. Pang, X.; Zhitomirsky, I. Electrophoretic deposition of composite hydroxyapatite-chitosan coatings. *Mater. Charact.* **2007**, *58*, 339–348. [CrossRef]
30. Lisuzzo, L.; Cavallaro, G.; Milioto, S.; Lazzara, G. Halloysite nanotubes coated by chitosan for the controlled release of Khellin. *Polymers* **2020**, *12*, 1766. [CrossRef]
31. Rapacz-Kmita, A.; Ewa, S.-Z.; Ziąbka, M.; Różycka, A.; Dudek, M. Instrumental characterization of the smectite clay–gentamicin hybrids. *Bull. Mater. Sci.* **2015**, *38*, 1069–1078. [CrossRef]
32. Raoufi, D. Synthesis and microstructural properties of ZnO nanoparticles prepared by precipitation method. *Renew. Energy* **2013**, *50*, 932–937. [CrossRef]
33. Sakka, S.; Coulthard, P. Implant failure: Etiology and complications. *Medicina Oral Patología Oral y Cirugia Bucal* **2011**, *16*, e42–e44. [CrossRef] [PubMed]
34. Flemming, H.C.; Wingender, J. The biofilm matrix. *Nat. Rev. Microbiol.* **2010**, *8*, 623–633. [CrossRef] [PubMed]
35. Jayaramudu, T.; Varaprasad, K.; Pyarasani, R.D.; Reddy, K.K.; Kumar, K.D.; Akbari-Fakhrabadi, A.; Mangalaraja, R.; Amalraj, J. Chitosan capped copper oxide/copper nanoparticles encapsulated microbial resistant nanocomposite films. *Int. J. Biol. Macromol.* **2019**, *128*, 499–508. [CrossRef]
36. Mishra, S.K.; Ferreira, J.M.F.; Kannan, S. Mechanically stable antimicrobial chitosan–PVA–silver nanocomposite coatings deposited on titanium implants. *Carbohydr. Polym.* **2015**, *121*, 37–48. [CrossRef] [PubMed]
37. Anaya-Esparza, L.M.; Ruvalcaba-Gómez, J.M.; Maytorena-Verdugo, C.I.; González-Silva, N.; Romero-Toledo, R.; Aguilera-Aguirre, S.; Pérez-Larios, A.; Montalvo-González, E. Chitosan-TiO_2: A versatile hybrid composite. *Materials* **2020**, *13*, 811. [CrossRef] [PubMed]
38. Al-Naamani, L.; Dobretsov, S.; Dutta, J. Chitosan-zinc oxide nanoparticle composite coating for active food packaging applications. *Innov. Food Sci. Emerg. Technol.* **2016**, *38*, 231–237. [CrossRef]
39. Silva, S.; Henriques, M.; Martins, A.; Oliveira, R.; Williams, D.W.; Azeredo, J. Biofilms of non-Candida albicans Candidaspecies: Quantification, structure and matrix composition. *Med. Mycol.* **2009**, *47*, 681–689. [CrossRef]
40. Prasad, A.S. Zinc in human health: effect of zinc on immune Cells. *Mol. Med.* **2008**, *14*, 353–357. [CrossRef]
41. Prasad, A.S.; Beck, F.W.; Endre, L.; Handschu, W.; Kukuruga, M.; Kumar, G. Zinc deficiency affects cell cycle and deoxythymidine kinase gene expression in HUT-78 cells. *J. Lab. Clin. Med.* **1996**, *128*, 51–60. [CrossRef]
42. Luna, S.M.; Silva, S.S.; Gomes, M.E.; Mano, J.F.; Reis, R.L. Cell adhesion and proliferation onto chitosan-based membranes treated by plasma surface modification. *J. Biomater. Appl.* **2010**, *26*, 101–116. [CrossRef] [PubMed]
43. Hu, H.; Zhang, W.; Qiao, Y.; Jiang, X.; Liu, X.; Ding, C. Antibacterial activity and increased bone marrow stem cell functions of Zn-incorporated TiO_2 coatings on titanium. *Acta Biomater.* **2012**, *8*, 904–915. [CrossRef] [PubMed]

44. Jin, G.; Cao, H.; Qiao, Y.; Meng, F.; Zhu, H.; Liu, X. Osteogenic activity and antibacterial effect of zinc ion implanted titanium. *Colloids Surf. B Biointerfaces* **2014**, *117*, 158–165. [CrossRef] [PubMed]
45. Yu, Y.; Jin, G.; Xue, Y.; Wang, N.; Liu, X.; Sun, J. Multifunctions of dual Zn/Mg ion co-implanted titanium on osteogenesis, angiogenesis and bacteria inhibition for dental implants. *Acta Biomater.* **2017**, *49*, 590–603. [CrossRef]
46. O'Connor, J.P.; Kanjilal, D.; Teitelbaum, M.; Lin, S.S.; Cottrell, J. Zinc as a Therapeutic Agent in Bone Regeneration. *Materials* **2020**, *13*, 2211. [CrossRef]
47. Qiao, Y.; Zhang, W.; Tian, P.; Meng, F.; Zhu, H.; Jiang, X.; Liu, X.; Chu, P.K. Stimulation of bone growth following zinc incorporation into biomaterials. *Biomaterials* **2014**, *35*, 6882–6897. [CrossRef]
48. Yu, J.; Xu, L.; Li, K.; Xie, N.; Xi, Y.; Wang, Y.; Zheng, X.; Chen, X.; Wang, M.; Ye, X. Zinc-modified calcium silicate coatings promote osteogenic differentiation through TGF-β/Smad pathway and osseointegration in osteopenic rabbits. *Sci. Rep.* **2017**, *7*, 3440. [CrossRef]
49. He, J.; Zhang, B.; Shao, L.; Feng, W.; Jiang, L.; Zhao, B. Biomechanical and histological studies of the effects of active zinc-coated implants by plasma electrolytic oxidation method on osseointegration in rabbit osteoporotic jaw. *Surf. Coat. Technol.* **2020**, *396*, 125848. [CrossRef]

coatings

Article

Modulus, Strength and Cytotoxicity of PMMA-Silica Nanocomposites

Sebastian Balos [1,*], Tatjana Puskar [2], Michal Potran [2], Bojana Milekic [2], Daniela Djurovic Koprivica [2], Jovana Laban Terzija [2] and Ivana Gusic [2]

1 Faculty of Technical Sciences, University of Novi Sad, 6 Trg Dositeja Obradovica, 21000 Novi Sad, Serbia
2 Faculty of Medicine, University of Novi Sad, 3 Hajduk Veljkova, 21000 Novi Sad, Serbia; tpuskar@uns.ac.rs (T.P.); michalpotran@gmail.com (M.P.); bojana.milekic@mf.uns.ac.rs (B.M.); daniela.djurovic-koprivica@mf.uns.ac.rs (D.D.K.); jovana.laban-terzija@mf.uns.ac.rs (J.L.T.); ivana.gusic@mf.uns.ac.rs (I.G.)
* Correspondence: sebab@uns.ac.rs; Tel.: +381-21-485-2339

Received: 12 May 2020; Accepted: 15 June 2020; Published: 23 June 2020

Abstract: Key advantages of Poly(methyl methacrylate)—PMMA for denture application are related to aesthetics and biocompatibility, while its main deficiency is related to mechanical properties. To address this issue, SiO_2 nanoparticle reinforcement was proposed, containing 0 to 5% nanosilica, to form nanocomposite materials. Flexural strengths and elastic moduli were determined and correlated to nominal nanoparticle content and zeta potential of the liquid phase nanoparticle solutions. Another issue is the biocompatibility, which was determined in terms of cytotoxicity, using L929 and MRC5 cell lines. The addition of nanoparticle was proved to be beneficial for increasing flexural strength and modulus, causing a significant increase in both strength and moduli. On the other hand, the formation of agglomerates was noted, particularly at higher nanoparticle loadings, affecting mechanical properties. The addition of nanosilica had an adverse effect on the cytotoxicity, increasing it above the level present in unmodified specimens. Cytotoxic potential was on the acceptable level for specimens with up to 2% nanosilica. Consequently, nanosilica proved to be an effective and biocompatible means of increasing the resistance of dental materials.

Keywords: nanocomposite; mechanical properties; cytotoxicity; nanosilica

1. Introduction

Poly(methyl methacrylate) (PMMA) is the most frequently used material for dentures today. Advantages of PMMA over previously used materials [1] include its aesthetics and biocompatibility. However, PMMA mechanical properties, particularly the strength and ductility are not optimal, leaving wide opportunities for further improvements. Indeed, the conducted studies found that over a half of them fail in the first three years [2]. Having in mind that dentures are mainly used by elderly people, often having limitations in mobility, the repair of fractured dentures presents an inconvenience and a significant problem. Therefore, increasing mechanical properties of denture material is of great significance, without compromising aesthetics and biocompatibility [3].

There are a number of ways to improve mechanical properties of PMMA materials, some being technological aftertreatments, while some used additional materials to improve properties of PMMA, creating composites or nanocomposites. A very attractive way of improving PMMA mechanical response is the application of microwave aftertreatment and water bath post-polymerization. These aftertreatments were the topic studied in the works by Vergani et al. [4], Balos et al. [5,6] and Urban et al. [7]. In these studies, it was found that PMMA mechanical properties can be increased by treatments after the curing was finished. Heating, by means of hot water bath or microwave irradiation proved to be effective, particularly for increasing mechanical properties of autopolymerized PMMA with a higher amount of unconverted monomer. By applying these technologies influence the marked reduction in residual monomer. Namely, the most significant reason why PMMA with residual monomer does not have convenient mechanical properties is that residual monomer is an unconverted polymer material. This acts as empty space, with the effect of a microvoid. As such, under load, it acts as a stress concentrator, actually weakening the material [8,9]. Another issue is the gum irritation and possibly even an allergic reaction to the pockets of monomer present in the PMMA [10].

Another opportunity for hot-cured PMMA, containing initially a lower amount of unconverted monomer mechanical properties can be increased by fibers or nanoparticle addition, effectively turning PMMA into a PMMA—based composite or a nanocomposite material. Carlos and Harrison [11] introduced ultra-high molecular weight polyethylene (UHMWPE) fibers into PMMA acrylic resin denture base material. However, some mechanical properties, such as hardness and impact strength were reduced compared to the PMMA matrix. Furthermore, the introduction of hydroxyapatite in studies by Chow et al. [12,13] of up to 5 wt.% can have variable effect on mechanical properties. It was shown that it can influence the increase in fracture toughness, however, flexural strength decreased.

The most likely explanation for such behavior is the formation of agglomerates that formed at these relatively high nanoparticle loadings. Other types of particles were tested, such as ZrO_2 and Al_2O_3 in the studies of Ayad et al. [14], Ellakwa et al. [15] and Alhareb and Ahmad [16]. Ayad et al. [14] reported that the relatively high loading of 15% ZrO_2 nano particles can influence the significant improvement in impact strength of PMMA. On the other hand, Ellakwa et al. [15] successfully applied Al_2O_3 nanoparticles in order to statistically significantly increase the flexural strength of PMMA. Finally, an interesting approach was applied by Alhareb and Ahmad [16], who reinforced PMMA with various ratios of Al_2O_3/ZrO_2 nanoparticles. It was shown that nanoparticles have a beneficial effect on increasing the flexural modulus, strength and fracture toughness. On the other hand, tensile modulus and strength were reduced. Combining different types of nanoparticles may be beneficial for reducing the tendency to form agglomerates and therefore be more effective in increasing mechanical properties. This is in accordance with the research by Efimov et al. [17].

One of the most widely used nanoparticle type in scientific community is nanosilica (nano-SiO_2). Several studies addressed the addition of nanosilica particles to PMMA resins. Yang et al. [18] and Balos et al. [19,20] found that nanosilica addition can have a significant influence on flexural strength, modulus, microhardness and fracture toughness. Relatively low nanoparticle loadings of up to 2% proved to be highly effective due to the formation of agglomerates at higher loadings, causing a non-homogenous distribution of nanoparticles, along with their polymer chain immobilization effect through the material section. Furthermore, relatively weak Van der Waals forces between nanoparticles in agglomerates cause the agglomerates to fracture under load, causing unstable crack propagation in the material, weakening it [21]. Nanoparticle addition may affect the materials biocompatibility, that is, the level of cytotoxicity. Namely, the addition of nanoparticles inevitably modifies the material structure, potentially affecting the nanocomposites biocompatibility. In the work by Yang and Nelson [22], PMMA/nanosilica composites were tested in tensile strengths and nanoparticle addition was proved to be beneficial in increasing the tensile strengths of over 50% and nearly doubling the modulus, while maintaining elongation at break.

In this study, PMMA denture reline resins were modified with nanosilica treated with hexamethyldisilazane (HMDS) providing hydrophobic properties. The selected increments of nanosilica content between individual specimens were relatively narrow, resulting in 12 nanosilica concentration overall, aimed at determining cytotoxicity threshold, a novelty by itself, aimed at detecting the influence of both SiO_2 and HMDS effect. Cytotoxicity tests were done on both mouse and human lung fibroblasts, with aim to find the more sensitive fibroblast type for these particular experimental conditions. Also, trends in flexural strengths and moduli values were found and correlated to the true particle size in the liquid phase. As such, this experimental setup was envisaged to give a comprehensive answer to the nanosilica effect on PMMA in both cytotoxic and mechanical response.

2. Materials and Methods

2.1. Materials

The base material used in this study is a two component PMMA, common in fabrication of acrylic dentures. This base material consisted of two components, the powder and the liquid. The powder had the composition consisting of pre-polymerized PMMA, pigments to emulate gums and benzoil peroxide (chemical formula $C_{14}H_{10}O_4$). The liquid contained methyl-methacrylate (MMA) and 3,4-Ethylenedioxy-*N*-methylamphetamine (EDMA). When mixed together with the powder, radical polymerization occurs, forming the bulk material.

Modification agent was nanosilica (Evonik Aerosil R812S), having the nominal particle size of 7 nm. These particles were pre-modified with a surface layer of hexamethyldisilazane (HMDS) to provide hydrophobic properties and be able to be mixed with the liquid phase. The nominal surface area of nanoparticles was 220 ± 25 m^2/g [23]. These nanoparticles were added to the liquid component and subsequently, in a mixed form was mixed together with the powder. The amount of nanoparticles was added so that the content of nanoparticles in the prepared specimens was 0.02; 0.05; 0.1; 0.2; 0.5; 0.7; 1; 1.5; 2; 2.5; 3 and 5 weight % (wt.%). Also, 0% nanoparticles were used to fabricate control specimens, to assess the effect of the nanoparticle reinforcement. Nanoparticles were weighed by the Ohaus Adventurer Pro analytic balance, accurate to 0.0001 g, while the subsequent mixing was done in two stages: by the IKA C-MAG HS7 magnetic stirrer and the Vevor PS-20A ultrasonic bath. To find the true size of particles in the liquid component in the modified specimens, the Malvern Instruments Zetasizer Nano ZS device was applied.

2.2. Specimen Fabrication

The liquid component (unmodified and modified with nanoparticles) was then mixed with the powder with a spatula and left to rest for 10 min in a closed glass cup. When the dough-state of the mixture was reached, the material was poured in isolated flask halves at 40 °C, pre-filled with hardened plaster. The mold was made by boiling out 50 mm × 50 mm × 4 mm wax models. Subsequently, the flasks were closed and the whole set was pressed at the pressure of 80 bars and fixed by clamping.

Curing was done in boiling water for 45 min to reduce the residual monomer content. After cooling to room temperature in cold water, cutting was done by the Struers Discotom apparatus, the standard metallographic abrasive cutter, with subsequent finishing by grinding. Grinding was done on the Struers Knuth Rotor metallographic grinding device, with standard 1500 grit silicone carbide (SiC) abrasive papers. Final measurement to achieve 50 mm × 6.25 mm × 2.5 mm specimen size was done by the Hyundai Altraco micrometer accurate to 0.01 mm, to make the specimens comply with ISO-178 standard [24].

2.3. Characterization of Mechanical Properties

To determine the effect of the nanoparticle addition, flexural strength and modulus were determined. Flexural testing was conducted by means of a three-point bend testing procedure, by using the Toyoseiki AT-L-118B universal mechanical tensile testing machine. This device was equipped with a three-point bending tool. Crosshead speed was set at 1 mm/min, while flexural strength and modulus were calculated in accordance to Equations (1) and (2), commonly used in flexural testing of polymer, composite and nanocomposite specimens.

$$\sigma = 3F_{max}L/(2BH^2) \tag{1}$$

$$E = FL^3/(4BH^3d) \tag{2}$$

where: σ is flexural strength of the material [MPa], Fmax is maximum load measured during flexural testing [N], L the distance between the flexural supports [mm], B and H are width and height of the samples [mm], E is the modulus of elasticity [GPa], while d is the deflection [mm] that corresponds to the loading F. Flexural test was conducted in accordance to the common ISO-178 standard [24]. Overall, tests were conducted on the basis of 10 specimens, used for flexural modulus and subsequently flexural strength testing. In this paper, average values were reported, as well as corresponding standard deviations.

One-way analysis of variance (ANOVA) with Tukey's post-hoc test was done to assess the mechanical test results and find the mathematical significance between specimens modified with different nanoparticle loadings between each other and versus the control specimen. The significance level selected was set at the common $\alpha = 0.05$. Statistical analysis was done by applying the Minitab 19 software.

2.4. SEM and EDX Characterization

Fracture surfaces were examined by the JSM-6460LV scanning electron microscope (SEM) (JEOL, Tokyo, Japan), to give answers to possibly different fracture mode. Furthermore, even more important, it was aimed at finding crack propagation data. This would indicate the partial strength of the polymerized material versus pre-polymerized powder that form the tested nanocomposite material. Before the SEM analysis, the specimens were coated with gold using the SCD-005 device (Bal-Tec/Leica, Wetzlar, Germany). Energy-dispersive X-ray sepectroscopy (EDS) with INCA Microanalysis system was conducted to examine agglomerated nanoparticles and differentiate them from the polymer in the background.

2.5. Cytotoxicity Testing

Biocompatibility was determined by two classes of fibroblasts: L929 mouse fibroblasts (American Type Culture Collection CCL1) and MRC-5 human fibroblasts (American Type Culture Collection CCL 171). The cells were grown in DMEM (Dulbeccos modified Eagles medium), supplemented with 10% of FCS (fetal calf serum), antibiotic and antimycotic solution. Subculturing the cells was performed twice a week. The fabrication of a single-cell suspension was performed by using 0.1% trypsin in Ethylenediaminetetraacetic acid (EDTA). These cell lines were cultured in 25 cm^2 flasks, in a 100% humid environment containing 5% CO_2. The dye exclusion test (DET) with tarpan blue [25] was applied to determine cell number and the amount of viable cells. The viability of cells was above 90%. Material samples were extracted in 4 mL of DMEM Sigma (Dulbecco's modified eagle medium) without serum after 3, 5, 7, and 21 days. Duration of cell incubation was 48 h. Growth inhibition was determined in accordance with tetrazolium colorimetric MTT procedure proposed by Mosmann [26].

The viable cells were seeded into a microtiter with 96 wells at seeding density of 5103 cells per well, assuring logarithmic growth throughout the test. Serial dilutions of the various eluates in 100 μL volumes were added for 48 h at 37 °C and with the addition of 5% CO_2. After the treatment described in this procedure, 10 μL of MTT solution sterilized by filtration (5 mg/ml) was added to the wells. Incubation was done for 3 h at the temperature of 37 °C. Afterwards, suction removal of the medium and MTT was conducted. The formazan products were solubilised in 100 μL 0.04 M HCl in isopropanol. Subsequently, the plates were tested on a Labsystem Multiscan MCC340 spectrophotometer by using the wavelength of 540/690 nm, with the wells containing only the complete medium and MTT acting as blank. The testing was repeated twice. Inhibition of growth was calculated in accordance with Equation (3).

$$\%K = (A_{test}/A_{control}) \times 100 \tag{3}$$

where A is the absorbance of test and control sample, respectively.

3. Results and Discussion

3.1. Zeta Sizer Results

The results of zeta sizer of the modified liquid component are shown in Figure 1. In all charts, for all concentrations of nanoparticles, the smallest detected particles are considerably larger than the nominal size of nanoparticles. That means that agglomeration occurs in all tested specimens, regardless of the nanoparticle content. These results suggest that the agglomeration occurs immediately after mixing of the nanoparticles and the liquid phase in spite of the presence of HMDS modification layer on the nanoparticles, magnetic stirring and ultrasonic bath treatment.

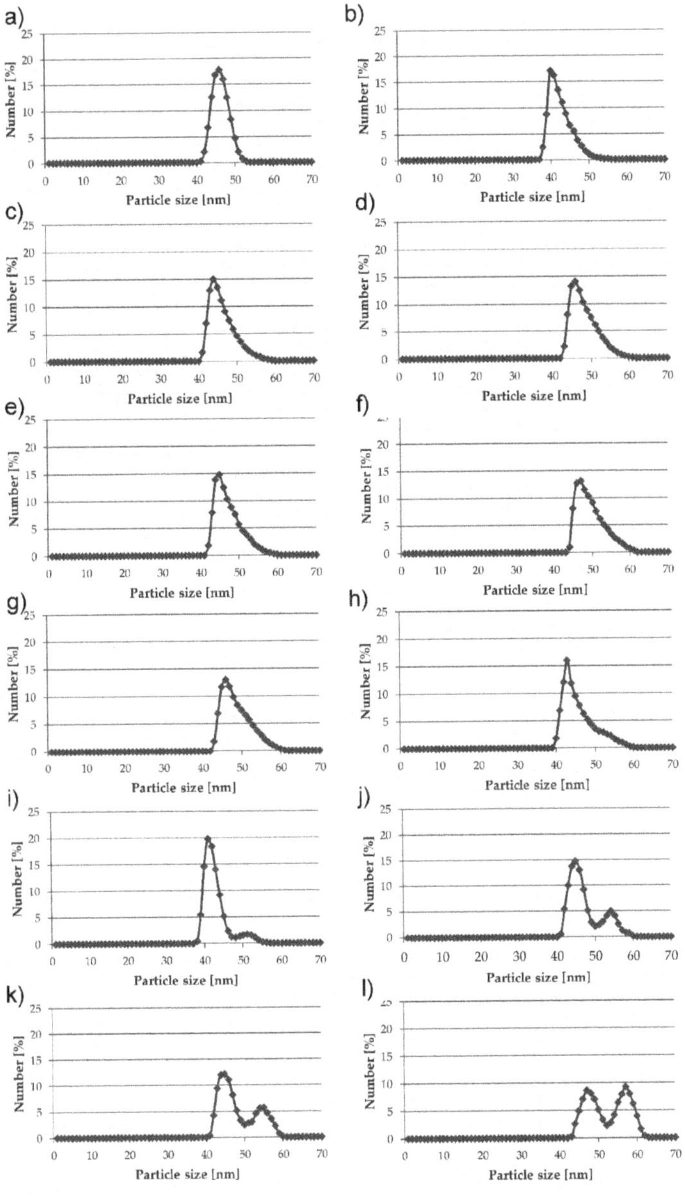

Figure 1. Zeta sizer results for modified liquid component that was subsequently mixed with powder, with a nanoparticle content of: (**a**) 0.02%; (**b**) 0.05%; (**c**) 0.1%; (**d**) 0.2%; (**e**) 0.5%; (**f**) 0.7%; (**g**) 1%; (**h**) 1.5%; (**i**) 2%; (**j**) 2.5%; (**k**) 3%; (**l**) 5%.

In Figure 1a–f, a single peak occurs with the maximum at 40–50 nm (specimens containing 0.02–0.7% nanoparticles). However, in Figure 1g,h, a transition occurs at the particle size of approximately 50 nm (1, 1.5%). At a higher nanoparticle content, a clear secondary peak occurs in the specimens modified with 2 and over 2% nanoparticles (Figure 1i–l). The secondary peak, which occurs beyond the particle size of 50 nm, becomes dominant in the specimen modified with 5% nanoparticles. This phenomenon suggests that an increase of nanoparticle content in the liquid phase increases the appearance of agglomeration as well, causing the occurrence of larger agglomerates in the nanocomposite material.

3.2. Mechanical Properties

The influence of nanoparticle content on flexural strength and modulus, along with the letter indication of statistical similarity between specimens is shown in Figure 2. Regarding statistical analysis, specimen groups that are marked with different letters are statistically different. The lowest mechanical properties were obtained in unmodified specimens which was expected. The highest flexural strength and modulus were obtained with 0.05% and 2% nanoparticles. The trends in flexural strength and modulus are similar, indicating that the effect of nanoparticle addition is similar in all modified specimens. Another observation in both charts depicting flexural strength and modulus is the notable waviness of average values: after the unmodified specimen 0, strength and modulus rises in specimen modified with 0.02% nanosilica, reaching maximum at 0.05%, after that the values drop in 0.1 and 0.2, rising again in 0.5%, dropping in 0.7 and 1%, rise again in 1.5 and 2%, and finally gradually drop down in 2.5, 3 and 5%.

The lowest values of flexural strength and modulus were obtained with the highest two nanoparticle contents of 3 and 5%. Also, it can be seen that standard deviations of tested specimens differ considerably. The smallest standard deviations were obtained in the unmodified specimen. As the nanoparticle content is increased, standard deviations are increased as well. This can be the result of somewhat stochastic effect of nanoreinforcement and possibly some degree of non-homogenous distribution. Similar findings were reported in [19,20,27], although [27] was aimed at increasing the properties of flowable dental composite by nanosilica addition.

The results of flexural strength and modulus are well correlated to each other and they follow a similar wavy pattern as the nanoparticle content is increased. This can be explained by zeta sizer results, which indicate that smaller particles in the liquid component also result in higher strengthening effect. Ideally, the particles detected would influence the appearance of a sharp peak, situated at the nanoparticle size as small as possible. Such distribution would indicate a relatively low agglomeration that would be effective in increasing the mechanical properties of the PMMA. The main strengthening effect of nanoparticles is reflected through the immobilization of polymer chains in the vicinity of nanoparticles.

Usually, around a micron distance around nanoparticle is considered as realistic for the reduction of polymer chain immobilization [28,29]. However, if the dispersion of nanoparticles is not optimal, that is, if relatively large agglomerates are formed, leaving areas of material without embedded nanoparticles and therefore without the mentioned polymer chain immobilization effect, the strengthening effect is not equally distributed throughout the material. This leaves considerable areas in the material microstructure unreinforced, without nanoparticles and without polymer chain immobilization. That means, there is a possibility for the crack to propagate between the reinforced fields, that is, through the unreinforced material, making the propagation easier.

After the specimen modified with 0.05% nanoparticles, the second specimen having the highest flexural strength and modulus is the specimen containing 2% nanoparticles, which is the second peak in the wavy trend of mechanical properties, Figure 2. This can be explained by the zeta sizer results, depicted in Figure 1. In Figure 1, the specimen modified with 2% of nanoparticles, a relatively sharp peak indicating around 20% of particles in the liquid phase has the size of 41 nm and there is also a relatively weak secondary peak at around 50 nm particle size, Figure 1i. This distribution is superior to both shown in Figure 1h (1.5% nanoparticles; 16% maximum at 44 nm) and Figure 1j

(2.5% nanoparticles; maximum at 45 nm and clear secondary peak at 54 nm). A more convenient particle distribution together with a higher nanoparticle content provides the specimen containing 2% nanoparticles with a higher flexural strength and modulus compared to the specimen containing 1.5% nanoparticles. A higher nanoparticle contents, 3 and 5%, provide lower mechanical properties due to a less convenient zeta potential shown in Figure 1k,l.

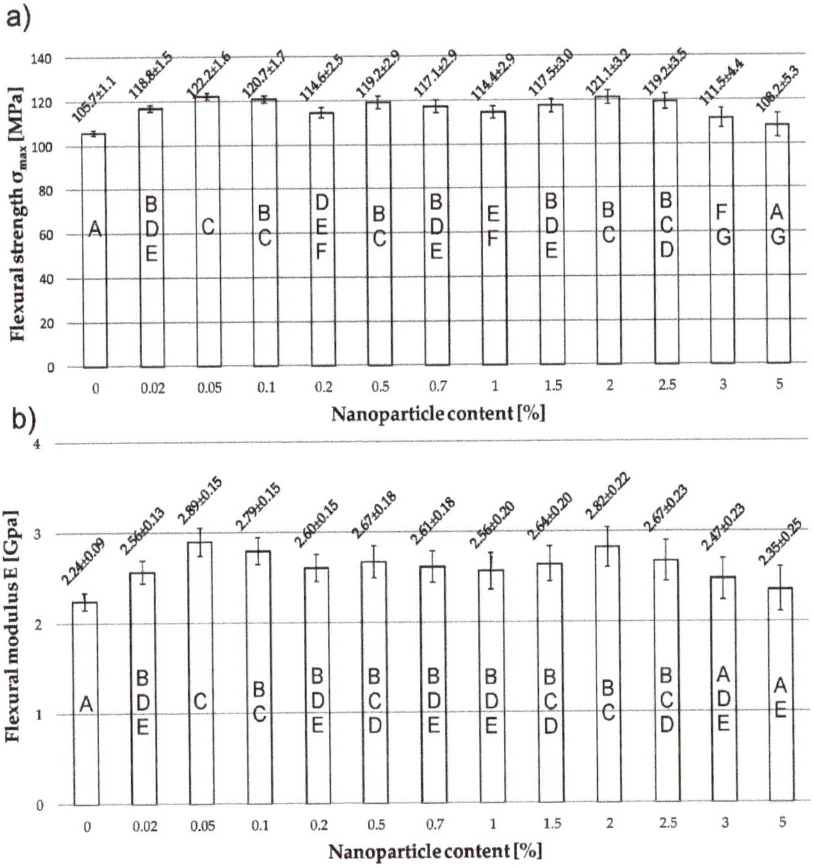

Figure 2. The nanoparticle content influence on flexural strength (**a**) and flexural modulus (**b**) of the unmodified and modified PMMA specimens. Letters indicate the statistical difference between specimens, whereby common letters indicate statistically insignificant differences.

3.3. Statistical Analysis

Statistical analysis results are shown in Figure 1, while in Tables 1 and 2, statistical analysis parameters are presented. These analyses confirmed that the majority of modified specimens exhibited statistically significant increase in flexural strength and modulus, mathematically showing the benefits from adding nanoparticles. The exception is the specimen modified with 5% nanosilica, which exhibits statistically similar flexural strength and modulus as the unmodified specimen. As such, the increase in mechanical properties is similar to literature data relating to similar PMMA based nanocomposites in flexural strength and modulus, as well as microhardness and fracture toughness [19,20,30].

Table 1. Analysis of variance for flexural strength.

Source	DF	Adj SS	Adj MS	F-Value	*p*-Value
Factor	12	3010	250.850	27.42	0.000
Error	117	1070	9.148		
Total	129	4081			

Table 2. Analysis of variance for flexural modulus.

Source	DF	Adj SS	Adj MS	F-Value	*p*-Value
Factor	12	3.942	0.32848	9.49	0.000
Error	117	4.048	0.03460		
Total	129	7.990			

3.4. SEM Analysis of Fracture Surfaces

SEM images of the fracture surfaces of the specimens modified with 0 and 2% nanoparticles, obtained after bend testing, are shown in Figure 3. Dark circular phases scattered through the cross section are primary PMMA powder particles of 30–100 μm approximate size, actually the powder component of the PMMA material. Between them, lighter in colour is the polymerized material, initially being the liquid MMA, which was modified with 7 nm nanoparticles. This light grey phase is effectively a nanocomposite material in modified specimens or PMMA in the control specimen (specimen 0). Both specimens (0 and 2% nanoparticle) exhibited the fracture that was brittle in nature, which was expected. Its nature is clear after assessing the typical river marks present in both specimens, which are more pronounced in fracture surface of the specimen 0.

However, a more significant differences between the specimen without nanoparticles (Figure 3a) and the modified specimen (Figure 3b) can be observed. Namely, in specimen 0 (Figure 3a), a crack exhibits a mixed path regarding the propagation during fracture. The crack in specimen 0 propagates both through and around powder particles, as well as through the polymerized material. That means that there is an issue of powder-polymerized material bond, also indicating that there is a significant difference in strength between the polymerized lighter area in colour and powder spheres.

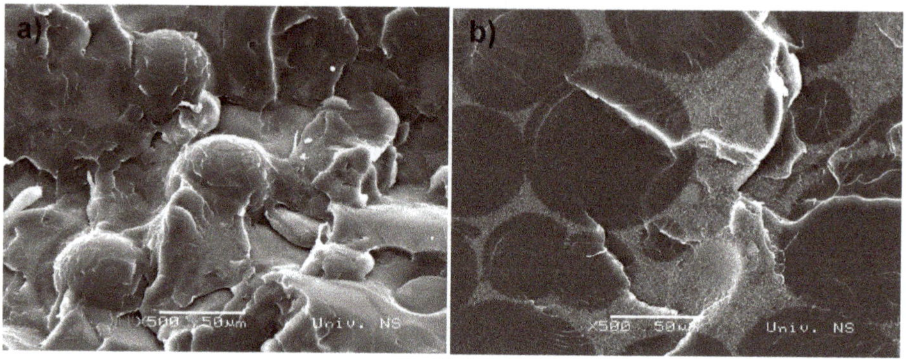

Figure 3. Fracture surfaces obtained after flexural testing of specimen 0 (**a**) and specimen modified with 2% nanosilica. A different crack propagation mode is revealed: predominantly through the nanocomposite between pre-polymerized powder particles (**a**) and through both pre-polymerized powder particles and polymerized nanosilica reinforced composite material (**b**).

A similar fracture morphology was found in autopolymerized PMMA [19,20], however, in this material, the crack propagates only through the polymerized light area. That means, the difference between the strength of the powder component and the polymerized PMMA around the powder is even higher. This is due to the fact that in autopolymerized material, the radical polymerization process is much quicker, leaving less time for mixing and homogenization. The second effect is the lack of pressing of the material, which eliminates a significant portion of porosity. Finally, in hot polymerization, versus autopolymerization, that is also called cold polymerization, a higher amount of unconverted monomer is present, which does not contribute to mechanical properties. Quite contrary, the unconverted monomer acts as a microvoid representing a crack initiation site weakening the material. In the specimen modified with 2% nanosilica, a crack propagates through the PMMA powder and the polymerized nanocomposite, Figure 3b. This indicates a significantly stronger bond between the powder and polymerized nanocomposite around it, suggesting their strengths are similar to each other.

3.5. EDX Analysis of Agglomerates

EDX analysis of the certain phases observed on the fracture surface of the specimen modified with 2% is shown in Figure 4, along with the spectrum obtained in Table 3. Clusters of agglomerates were observed on the fracture surface, Figure 4. This indicates that agglomerates tend to cluster themselves together during fabrication process of the modified PMMA. On the other hand, it can be observed that these clusters consist of agglomerates and that these agglomerates have a similar size to those found by zeta sizer for this specimen modified with 2% nanosilica.

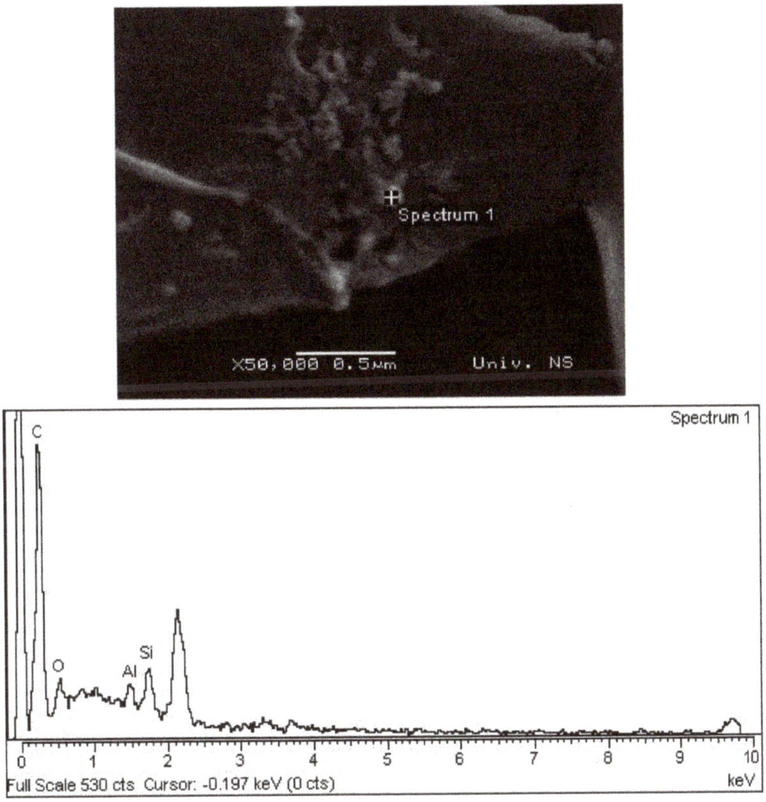

Figure 4. SEM images of fracture surface with a cluster of agglomerates (specimen modified with 2% nanoparticles) and EDX spectrum of particles. The analyzed spectrum contains Si and O, indicating that the point of interest is SiO_2 cluster of particles.

Table 3. Quantitative results of EDX analysis in wt.%.

Element	wt.%
C	73.07
O	22.11
Al	1.28
Si	3.54

EDX analysis reveals that this area contains silicon and oxygen, indicators of SiO_2. Although the content of Silicon is relatively low, EDX analysis analyses the material in depth, so a relatively large amount of PMMA is analyzed.

3.6. The Application of the Rule of Mixtures

The rule of mixtures (ROM) and the inverse rule of mixtures (IROM) are two common ways to predict the properties of composite materials in general, as shown in Equations (4) and (5) respectively [31]:

$$E_{ROM} = f_m E_m + f_r E_r \tag{4}$$

$$1/E_{IROM} = f_m/E_m + f_r/E_r \tag{5}$$

where E_{ROM} is the modulus of elasticity obtained by the ROM; E_{IROM} is the modulus of elasticity obtained by the IROM [GPa]; f_m is the volume fraction of the matrix; E_m is the modulus of elasticity of the matrix; f_r is the volume fraction of the reinforcement and E_r is the modulus of elasticity of the reinforcement.

The effectiveness of nanoparticle reinforcement can also be assessed by ROM and IROM. Modulus of elasticity of 71 GPA for SiO_2 was used for nanoparticles [32], while the unmodified PMMA (0% nanoparticles) was used for the matrix.

The calculated theoretical rule of mixtures and inverse rule of mixtures, as well as experimentally obtained values in this study are presented in Table 4. Table 4 reveals that elastic moduli of the specimens modified with 0.02, 0.05, 0.1, 0.2, 0.5 and 0.7% nanoparticles are higher than the moduli calculated in accordance to the ROM, with the study to rule of mixtures ratio higher than 1. The maximum is reached at 0.05% nanoparticle content, which is also the highest value obtained in this study.

Table 4. Theoretical values calculated based on rule of mixtures (ROM) and inverse rule of mixtures (IROM) compared to average elastic moduli obtained in this work.

	Modulus [GPa] for the Specimen Containing the Following Amount of Nanosilica [%]											
	0.02	0.05	0.1	0.2	0.5	0.7	1	1.5	2	2.5	3	5
ROM	2.25	2.26	2.27	2.30	2.40	2.46	2.56	2.72	2.87	3.03	3.19	3,82
IROM	2.24	2.24	2.24	2.24	2.25	2.25	2.26	2.26	2.26	2.27	2.27	2.29
This study	2.56	2.89	2.79	2.60	2.67	2.61	2.56	2.64	2.82	2.67	2.47	2.35
This study/ROM ratio	1.14	1.28	1.23	1.13	1.11	1.06	1.00	0.97	0.98	0.88	0.77	0.61

As the specimen nanoparticle content increases, the ratio drops, reaching the minimum at 5% nanoparticle content (ratio of 0.61). Furthermore, in the specimens having the highest concentrations of nanoparticles (3 and 5%), nanoparticles behave as particulate composite material fillers, since the moduli values approach those obtained by the IROM [31].

3.7. Results of Cytotoxicity

The influence of nanoparticle content and days of elution of the viability of L929 mouse fibroblasts and MRC 5 human lung fibroblasts is presented in Figure 4. Viabilities are the highest in specimens after the shortest time of elution. The lowest cytotoxicity (the highest viability) was obtained in unmodified specimens (0%), while standard deviations are the lowest in these specimens as well. This is a similar trend to mechanical properties, where standard deviations were obtained in the specimens containing nanoparticles.

If the results obtained with two different cell lines used in this study are compared, it can be seen that trends are similar. However, it can be observed that MRC human lung fibroblasts are slightly more sensitive than L929 mouse fibroblasts, which is due to differences that exist between species. Results gained from human cell culture give more relevant results.

Biocompatibility tests based on mouse and human lung fibroblasts reveal a notable increase in cytotoxicity as the amount of nanoparticles in specimens increase. On the basis of presented results in Figure 5 and relevant standards [33,34], the nanoparticle content of up to 2% can be considered non-cytotoxic, as the cell viability reduction is 30% or less. That means that increasing nanoparticle content is both cytotoxic and non-beneficial from the point of view of mechanical properties.

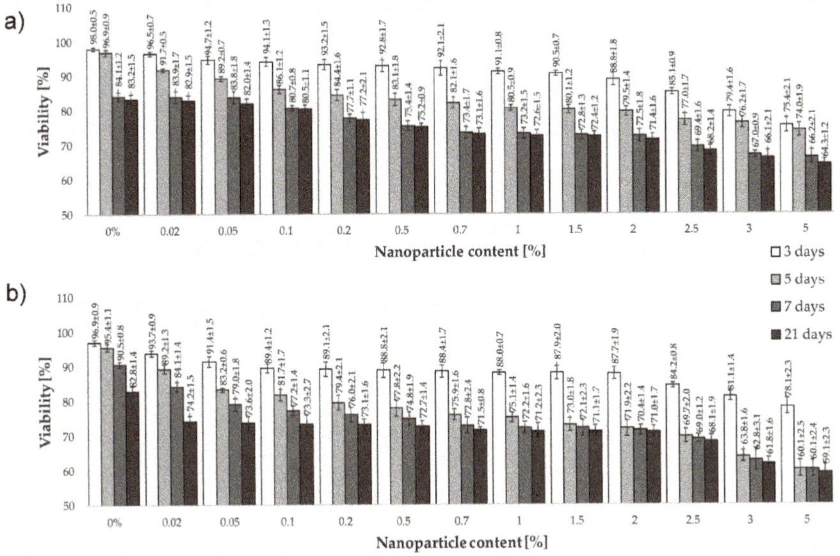

Figure 5. The influence of nanoparticle content and days of elution on the viability of L929 mouse fibroblasts (**a**) and MRC 5 human lung fibroblasts (**b**).

This proves to be highly selective, leaving 0.05% nanoparticle specimens as the optimal of all those tested in this study.

4. Conclusions

In this study, PMMA denture reline resins were modified with nanosilica treated with hexamethyldisilazane (HMDS) providing hydrophobic properties were tested in flexural strength and modulus, as well as cytotoxicity. Based on presented results and limitations in this study, the following conclusions can be established:

- The addition of nanoparticles is beneficial to an increased flexural strength and modulus of elasticity of hot polymerized PMMA.
- The increase in flexural strengths is statistically significant in relation to control group of specimens, except for specimens reinforced with 3% nanoparticles.
- In elastic moduli, all specimens exhibited statistically significant increase, except of the specimens reinforced with 3 and 5% nanoparticles.

- Mechanical properties are in good agreement with zeta sizer results, with the highest strength and modulus closely corresponding to the highest amount of smaller particles in the liquid phase.
- The size of detected particles in all modified specimens is considerably larger than the nominal nanoparticle size, indicating agglomeration occurred in all specimens.
- The highest flexural strength and modulus were obtained with the addition of 0.05% of nanoparticles.
- The addition of nanoparticles increases standard deviations of flexural strength and modulus.
- Specimens suffered a brittle fracture mode.
- The most efficient nanoparticle addition is 0.05%, with the average value of moduli higher than those the rule of mixtures suggests.
- The highest nanoparticle contents of 3% and 5% behave like fillers to the polymer material, providing the moduli that are closer to the inverse rule of mixtures.
- While an increase in nanoparticle content increases the material cytotoxicity, with specimens containing up to 2% of nanosilica it can be considered as non-cytotoxic.

It can be said that the most efficient and optimal 7 nm hydrophobic nanosilica content into hot polymerized PMMA is 0.05%. This loading provides the most convenient particle distribution in the liquid phase, the highest mechanical properties in terms of flexural strength and modulus, and can be considered as non-cytotoxic, by using both mouse and human lung fibroblasts.

Author Contributions: Conceptualization, S.B. and T.P.; investigation, M.P.; writing—original draft preparation, S.B.; writing—review and editing, D.D.K., T.P., and I.G., funding acquisition, B.M. and J.L.T. All authors have read and agreed to the published version of the manuscript.

Funding: This work is self-funded by the authors.

Acknowledgments: This research was supported by the Department of Production Engineering, Faculty of Technical Sciences Novi Sad Serbia, entitled "Modern technologies in materials science and welding technology".

Conflicts of Interest: The authors declare no conflict of interest.

References

1. Jerolimov, V. *Osnove Stomatoloških Materijala*; School of Dentistry, University of Zagreb, SFZG: Zagreb, Croatia, 2005; p. 80.
2. Hargreaves, A.S. The prevalence of fractured dentures: A survey. *Br. Dent. J.* **1969**, *126*, 451–455. [PubMed]
3. El-Sheikh, A.M.; Al-Zahrani, S.B. Causes of denture fracture: A survey. *Saudi Dent. J.* **2006**, *18*, 149–154.
4. Vergani, C.E.; Seo, R.S.; Pavarina, A.C.; Dos Santos Nunes Reis, J.M. Flexural strength of autopolymerizing denture reline resins with microwave postpolymerization treatment. *J. Prosthet. Dent.* **2005**, *93*, 577–583. [CrossRef] [PubMed]
5. Balos, S.; Balos, T.; Sidjanin, L.; Markovic, D.; Pilic, B.; Pavlicevic, J. Study of PMMA Biopolymer Properties Treated by Microwave Energy. *Mater. Plast.* **2011**, *48*, 127–131.
6. Balos, S.; Balos, T.; Sidjanin, L.; Markovic, D.; Pilic, M.; Pavlicevic, J. Flexural and Impact Strength of Microwave Treated Autopolymerlized Poly (Methyl-Methacrylate). *Mater. Plast.* **2009**, *46*, 261–265.
7. Urban, V.M.; Machado, A.L.; Oliveira, R.V.; Vergani, C.E.; Pavarina, A.C.; Cass, Q.B. Residual monomer of reline acrylic resins: Effect of water-bath and microwave post-polymerization treatments. *Dent. Mater.* **2007**, *23*, 363–368. [CrossRef]
8. Lee, S.Y.; Lai, Y.L.; Hsu, T.S. Influence of polymerization conditions on monomer elution and microhardness of autopolymerized polymethyl methacrylate resin. *Eur. J. Oral Sci.* **2002**, *110*, 179–183. [CrossRef]
9. Fletcher, A.M.; Purnaveja, S.; Amin, W.M.; Ritchie, G.M.; Moradians, S.; Dodd, A.W. The level of residual monomer in self-curing denture-base materials. *J. Dent. Res.* **1983**, *62*, 118–120. [CrossRef]
10. Azzarri, M.J.; Cortizo, M.S.; Alessandrini, J.L. Effect of the curing conditions on the properties of an acrylic denture base resin microwave-polymerized. *J. Dent.* **2003**, *31*, 463–468. [CrossRef]
11. Carlos, N.B.; Harrison, A. The effect of untreated UHMWPE beads on some properties of acrylic resin denture base material. *J. Dent.* **1997**, *25*, 59–64. [CrossRef]

12. Chow, W.S.; Tay, H.K.; Azlan, A.; Ishak, Z.A.M. Mechanical and thermal properties of hydroxyaptite filled poly (methyl methacrylate) composites. In Proceedings of the Polymer Processing Society 24th Annual Meeting—PPS-24, Salerno, Italy, 15–19 June 2008.
13. Chow, W.S.; Khim, L.Y.A.; Azlan, A.; Ishak, Z.A.M. Flexural properties of hydroxyapatite reinforced poly(methyl methacrylate) composite. *J. Reinf. Plast. Comp.* **2008**, *27*, 945–952. [CrossRef]
14. Ayad, N.M.; Badawi, M.F.; Fatah, A.A. Effect of reinforcement of high-impact acrylic resin with zirconia on some physical and mechanical properties. *Arch. Oral Res.* **2008**, *4*, 145–151.
15. Ellakwa, A.E.; Morsy, M.A.; El-Sheikh, A.M. Effect of aluminum oxide addition on the flexural strength and thermal diffusivity of heat-polymerized acrylic resin. *J. Prosthodont.* **2008**, *17*, 439–444. [CrossRef] [PubMed]
16. Alhareb, A.O.; Ahmad, Z.A. Effect of Al_2O_3/ZrO_2 reinforcement on the mechanical properties of PMMA denture base. *J. Reinf. Plast. Comp.* **2011**, *30*, 86–93. [CrossRef]
17. Efimov, A.; Lizunova, A.; Sukharev, V.; Ivanov, V. Synthesis and Characterization of TiO_2, Cu_2O and Al_2O_3 Aerosol Nanoparticles Produced by the Multi-Spark Discharge Generator. *Korean J. Mater. Res.* **2016**, *26*, 123–129. [CrossRef]
18. Yang, H.; Wu, S.; Hu, J.; Wang, Z.; Wang, R.; Huiming, H. Influence of nano-ZrO_2 additive on the bending strength and fracture toughness of fluoro-silicic mica glass–ceramics. *Mater. Des.* **2011**, *32*, 1590–1593. [CrossRef]
19. Balos, S.; Pilic, B.; Markovic, D.; Pavlicevic, J.; Luzanin, O. Poly (methyl-methacrylate) nanocomposites with low silica addition. *J. Prosthet. Dent.* **2014**, *111*, 327–334. [CrossRef]
20. Balos, S.; Pilic, B.; Petrovic, D.; Petronijevic, B.; Sarcev, I. Flexural strength and modulus of autopolimerized poly (methyl methacrylate) with nanosilica. *Vojnosanit. Pregl.* **2018**, *75*, 564–569. [CrossRef]
21. Bera, O.; Pilic, B.; Pavlicevic, J.; Jovicic, M.; Hollo, B.; Mesaros Szecsenyi, K. Preparation and thermal properties of polystyrene/silica nanocomposites. *Thermochim. Acta* **2011**, *515*, 1–5. [CrossRef]
22. Yang, F.; Nelson, G.L. PMMA/silica nanocomposite studies: Synthesis and properties. *J. Appl. Polym. Sci.* **2004**, *91*, 3844–3850. [CrossRef]
23. Evonik Site. Available online: https://www.aerosil.com/product/aerosil/en/products/hydrophobic-fumed-silica/pages/default.aspx (accessed on 10 December 2019).
24. *ISO 178:2019 Plastics—Determination of Flexural Properties*; ISO: Geneva, Switzerland, 2019.
25. Bogdanović, G.; Raletić-Savić, J.; Marković, N. In Vitro Assays for Antitumor-Drug Screening on Human Tumor Cell Lines: Dye Exclucion Test and Colorimetric Cytotoxicity Assay. *Arch. Oncol.* **1994**, *2*, 181–184.
26. Mosmann, T. Rapid calorimetric assay for cellular growth and survival: Application to proliferation and cytotoxicity assays. *J. Immunol. Meth.* **1983**, *65*, 55–63. [CrossRef]
27. Balos, S.; Pilic, B.; Petronijevic, B.; Markovic, D.; Mirkovic, S.; Sarcev, I. Improving mechanical properties of flowable dental composite resin by adding silica nanoparticles. *Vojnosanit. Pregl.* **2013**, *70*, 477–483. [CrossRef] [PubMed]
28. Fragiadakis, D.; Pissis, P.; Bokobza, L. Glass transition and molecular dynamics in poly (dimethylsiloxane)/ silica nanocomposites. *Polymer* **2005**, *46*, 6001–6008. [CrossRef]
29. Bera, O.; Pavlicevic, J.; Jovicic, M.; Stoiljkovic, D.; Pilic, B.; Radicevic, R. The influence of nanosilica on styrene free radical polymerization kinetics. *Polym. Comp.* **2012**, *33*, 262–266. [CrossRef]
30. Elshereksi, N.W.; Mohamed, S.H.; Arifin, A.; Mohd Ishak, Z.A. Effect of filler incorporation on the fracture toughness properties of denture base poly (methyl methacrylate). *J. Phys. Sci.* **2009**, *20*, 1–12.
31. Smallman, R.E.; Bishop, R.J. *Modern Physical Metallurgy and Materials Engineering*; Butterworth Heinemann: Oxford, UK, 1999; pp. 321–362.
32. Smallman, R.; Ngan, A.H.W. *Physical Metallurgy and Advanced Materials*; Butterworth Heinemann: Oxford, UK, 2007; p. 566.
33. *ISO 7405:2018 Dentistry—Evaluation of Biocompatibility of Medical Devices Used in Dentistry*; ISO: Geneva, Switzerland, 2018.
34. ISO 10993-5. *Biological Evaluation of Medical Devices—Part 5: Tests for In Vitro Cytotoxicity*; International Organization for Standardization: Geneva, Switzerland, 2009.

coatings

Review

Protein–TiO_2: A Functional Hybrid Composite with Diversified Applications

Luis Miguel Anaya-Esparza [1,2], Zuamí Villagrán-de la Mora [3], Noé Rodríguez-Barajas [2], Teresa Sandoval-Contreras [1], Karla Nuño [4], David A. López-de la Mora [4,*], Alejandro Pérez-Larios [2,*] and Efigenia Montalvo-González [1,*]

1 Laboratorio Integral de Investigación en Alimentos, Tecnológico Nacional de México-Instituto Tecnológico de Tepic, Av. Tecnológico 2595 Fracc, Lagos del Country, Tepic 63175, Mexico; lumianayaes@ittepic.edu.mx (L.M.A.-E.); tesysval@gmail.com (T.S.-C.)
2 Laboratorio de Materiales, Agua y Energía, División de Ciencias Agropecuarias e Ingenierías, Centro Universitario de los Altos, Universidad de Guadalajara, Av. Rafael Casillas Aceves 1200, Tepatitlán de Morelos 47600, Mexico; noe.rbarajas@academicos.udg.mx
3 División de Ciencias Biomédicas, Centro Universitario de los Altos, Universidad de Guadalajara, Av. Rafael Casillas Aceves 1200, Tepatitlán de Morelos 47600, Mexico; blanca.villagran@academicos.udg.mx
4 División de Ciencias de la Salud, Centro Universitario de Tonalá, Universidad de Guadalajara, Av. 555 Ejido San José Tateposco, Nuevo Perif. Ote., Tonalá 475425, Mexico; karlajanette.nuno@cutonala.udg.mx
* Correspondence: david.ldelamora@academicos.udg.mx (D.A.L.-d.l.M.); alarios@cualtos.udg.mx (A.P.-L.); emontalvo@ittepic.edu.mx (E.M.-G.)

Received: 4 November 2020; Accepted: 2 December 2020; Published: 7 December 2020

Abstract: Functionalization of protein-based materials by incorporation of organic and inorganic compounds has emerged as an active research area due to their improved properties and diversified applications. The present review provides an overview of the functionalization of protein-based materials by incorporating TiO_2 nanoparticles. Their effects on technological (mechanical, thermal, adsorptive, gas-barrier, and water-related) and functional (antimicrobial, photodegradation, ultraviolet (UV)-protective, wound-healing, and biocompatibility) properties are also discussed. In general, protein–TiO_2 hybrid materials are biodegradable and exhibit improved tensile strength, elasticity, thermal stability, oxygen and water resistance in a TiO_2 concentration-dependent response. Nonetheless, they showed enhanced antimicrobial and UV-protective effects with good biocompatibility on different cell lines. The main applications of protein–TiO_2 are focused on the development of eco-friendly and active packaging materials, biomedical (tissue engineering, bone regeneration, biosensors, implantable human motion devices, and wound-healing membranes), food preservation (meat, fruits, and fish oil), pharmaceutical (empty capsule shell), environmental remediation (removal and degradation of diverse water pollutants), anti-corrosion, and textiles. According to the evidence, protein–TiO_2 hybrid composites exhibited potential applications; however, standardized protocols for their preparation are needed for industrial-scale implementation.

Keywords: proteins; titanium dioxide; functionalization; hybrid composites

1. Introduction

Nowadays, the development of eco-friendly materials with advanced characteristics and diverse applications is an active research area [1,2]. Hybrid compounds are composites that consist of combining inorganic–inorganic (e.g., TiO_2–Ag), organic–organic (e.g., wheat gluten–cellulose), and organic–inorganic (e.g., collagen–TiO_2) [3–5], and they can be synthesized by spin and dip coating, slot-casting, electrochemical self-assembly, and chemical vapor, atomic or molecular layer deposition methods [6]. In general, hybridization or functionalization of organic compounds by incorporating inorganic compounds is a strategy that enables the attainment of hybrid materials with beneficial properties and new functionalities [6,7]. Recently, titanium dioxide (TiO_2) has been used as a reinforcement agent to develop organic–inorganic hybrid materials with improved physicochemical, mechanical, UV- and gas-barrier, water resistance, and antimicrobial properties [1,3,8–11].

TiO_2 is an amphoteric, inert, non-toxic, biocompatible metal oxide that exhibited thermal and chemical stability for diverse applications with a relatively low cost of production [12]. The wide use of TiO_2 is to support its photocatalytic, adsorptive, UV-blocking, and antimicrobial properties [13–15]. It has been employed for environmental remediation in dye removal from aqueous media [16]. Moreover, TiO_2 is the main source of white pigments for food, pharmaceutical, and cosmetic applications in compliance with the recommended safe dosage [17–19]. Currently, there is a great interest in combining protein-based materials with inorganic compounds like titanium dioxide to fabricate protein–TiO_2 hybrid structures with improved physical and chemical properties, which open new opportunities and applications [1].

In the last decade, protein–TiO_2 hybrid composites and their potential applications have been explored [4,20–24]. Wang et al. [23] developed a soy protein isolate film combined with TiO_2 with antimicrobial properties against *Escherichia coli* and *Staphylococcus aureus*. Similar trends were reported when a marine algae protein-based film functionalized with TiO_2 was used [25]. Fathi et al. [16] informed that the sesame protein–TiO_2 hybrid film showed photocatalytic degradation of the methylene blue dye under UV-light radiation. Qingyan et al. [26] made gelatin film reinforced with TiO_2 with improved mechanical and UV-protective properties. Fan et al. [27] fabricated a collagen–chitosan–TiO_2 porous scaffold for wound-healing purposes. Meanwhile, He et al. [27] developed an active packaging with marine alga (*Gracilaria lemaneiformis*) protein isolate and TiO_2 for cherry tomatoes preservation, while a whey protein–TiO_2 hybrid film was used to prolong the shelf life of chilled and lamb meats [28,29]. Furthermore, the incorporation of TiO_2 in diverse protein-based materials (collagen, gelatin, soy, hey, marine alga, kefiran, zein, sesame, sodium caseinate, and wheat gluten) had a positive impact on the technological (mechanical, water resistance, and gas-barrier) and functional (antimicrobial and UV-protective) properties, which exhibited potential uses for various applications [16,17,20,24,25,30,31].

This review summarizes the advantages and limitations of protein-based material functionalization by adding TiO_2 nanoparticles, offers and provides an overview of their photocatalytic and antimicrobial properties, environmental remediation, potential food and non-food packaging, pharmaceutical, cosmetics, textile, and biomedical applications.

2. Proteins: Applications and Limitations

Proteins are biological molecules composed of α-amino acids connected by peptide bonds, which can be obtained from plant-derived or animal origins [32]. For example, zein and gluten are cereal proteins [4,33], meanwhile, other proteins can be obtained from legumes such as soy [19]. Collagen and gelatin are extracted from meat, fish, and poultry by-products [1,34], whereas whey protein is a by-product of dairy manufacturing [29]. They exhibited interesting technological and functional properties for diversified applications, associated with their composition and structure [2]. Moreover, they are non-toxic, abundant, readily available, biodegradable, low-cost, and biocompatible to combine with enzymes, microorganisms, and organic and inorganic compounds [6,32,35]. Most of the applications described in the literature for protein-based materials are focused on developing packaging materials for food and non-food purposes, or biomedical applications such as wound-healing materials, as shown in Table 1.

Table 1. Potential applications of some protein-based materials.

Protein Source	Application	Ref.
Yellow pea protein isolate	Food and non-food packaging	[36]
Whey protein	Food and non-food packaging	[37]
Corn zein	Food and non-food packaging	[38]
Soy protein isolate	Food and non-food packaging	[39]
Rice bran	Food and non-food packaging	[40]
Wheat gluten	Food and non-food packaging	[41]
Gelatin	Food and non-food packaging	[42]
Gelatin	Biomedical	[43]
Keratin	Biomedical	[44]

Acquah et al. [36] fabricated a yellow pea (*Pisum sativum*) protein-based film with potential food and non-food packaging applications. It exhibited moderate water solubility (36.5%), good mechanical properties (elongation of 65%, a tensile strength of 0.65 MPa, and elastic modulus of 6.65 MPa), as well as good thermal properties (glass transition of 95.5 °C), but high moisture uptake (82%) due to its hydrophilic nature (contact angle of 60°), affecting its quality as a packaging material. Agudelo-Cuartas et al. [37] mentioned that whey protein-based films showed great potential for packaging purposes (good mechanical properties); however, their high-water solubility (59%) and water vapor permeability (1.4×10^{-10} $g \cdot m^{-1} \cdot s^{-1} \cdot Pa^{-1}$) limit their uses in foods with high water content (e.g., meat). According to Guo et al. [38], the protein-based film's mechanical properties are influenced by storage conditions (temperature and relative humidity). They found that tensile strength and elongation at break of a zein film were negatively affected when relative humidity and temperature increased from 34% to 80% and from 5 to 35 °C, respectively. They argue that the available –SH groups in the protein structure decreased gradually during storage by water absorption, implying new and weak interactions.

Su et al. [39] reported that soy protein isolate film exhibited good biodegradability and gas-barrier properties against oxygen and carbon dioxide when relative humidity was low, which are suitable featuring for the development of packaging materials. Wang et al. [40] suggested that modification of protein structure by alkaline conditions could be an alternative to improve the technological properties of protein-based films. They reported that the formation of protein aggregates in a rice bran film treated at pH 11 improved their physical, mechanical, and thermal properties, associated with an increase in the β-sheet content and non-covalent interactions, due to the modification of the protein structure.

Additionally, gelatin-based films exhibited great potential for fabricating food packaging or wound-healing materials; however, due to their hygroscopic nature, they needed to be combined with a crosslinking or plasticizer agent (organic or inorganic) to improve their water resistance and thermal stability [42,43]. It has been reported that keratin films are too rigid, and the addition of glycerol improved their flexibility and mechanical resistance, which are suitable for biomedical applications [44]. Similar trends were reported in a wheat gluten film by adding glycerol, but its thermal stability was improved and could be used for packaging purposes [41].

In general, protein-based films exhibited great potential applications; however, their functionality depends on their molecular characteristics, complexity, superficial charge, denaturation tendency, water resistance, and thermal stability [35]. Therefore, the incorporation of organic and inorganic materials in the protein matrix is a viable strategy to enhance their functional and technological properties [29,32,45]. Table 2 shows some protein-based materials functionalized with organic and inorganic compounds to form hybrid composites with potential applications.

Table 2. Application of some functionalized protein-based materials.

Protein Source	Functional Agent	Application	Ref.
Gelatin	Silver-NPs	Active food packaging.	[46]
Gelatin	Resorcinol and silver-NPs	Active food packaging.	[47]
Caseinate/gelatin	Tannins	Active food packaging.	[48]
Sodium caseinate	ZnO-NPs and REO	Active food packaging.	[49]
Whey protein	Montmorillonite and citric acid	Active food packaging.	[50]
Whey protein	Organic acids and nisin	Active food packaging.	[51]
Furcellaran/whey protein	Yerba mate extracts	Active food packaging.	[52]
Yellow pea protein isolate	Whey protein isolate	Active food packaging.	[36]
Fish protein isolate	Gelatin and ZnO-NPs	Active food packaging.	[53]
Soy protein hydrolysate	Silica	Environmental remediation.	[54]
Soy protein isolate	Tragacanth, silica, and lycopene	Environmental remediation.	[55]
Silk fibroin	Ag NPs	Biomedical.	[56]
Egg white protein	Silk fibroin	Biomedical.	[57]

NPs: nanoparticles; ZnO: Zinc oxide; REO: rosemary essential oil.

According to the evidence, the incorporation of organic and inorganic compounds improves the technological (water and thermal resistance, mechanical, and adsorptive) and functional (antimicrobial activity and biocompatibility) properties of protein-based materials, associated with their ability to form intramolecular bonds through covalent and non-covalent interactions with the functional groups ($-NH_2$, $-OH$, $-COOH$, and $-SH$) of the protein structure [6,29].

Additionally, usage of TiO_2 as a functional agent to enhance the technological properties of diverse protein-based materials has been widely explored in the last years, mainly for the chemical and physical interactions between protein structure and TiO_2, which could be developed using diverse methodologies.

3. Possible Structural Interaction between R-Groups Amino Acid with TiO_2 Nanoparticles

A major understanding of the interactions between proteins and TiO_2 surfaces will be a potential core for many applications in bio-nanotechnology [58]. Ranjan et al. [59] in silico observed that the TiO_2 (1.09 nm) nanoparticles bind to 13 immunological proteins (Table 3), using a docking simulation program (AutoDock 4.0), a computed atlas of surface topography of proteins (CASTp) and PyMol software (version 1.5.0.4). They observed that nano-TiO_2 bound with a positively charged R-group (lysine, arginine, and histidine) and nonpolar aliphatic R-groups amino acid (proline, glycine, alanine, valine, leucine, methionine, and isoleucine) containing amino acids, most frequently with lysine and proline. On the other hand, TiO_2 had less affinity with the aromatic R-group (phenylalanine, tyrosine, and tryptophan), polar uncharged R-groups (serine, threonine, cysteine, asparagine, and glutamine), and negatively charged R-group (aspartate and glutamate)-containing amino acids. According to the authors, the affinity of TiO_2 with the amino acids depends on the ability to form stable hydrogen bonds, which depend on the binding and intermolecular energy of each amino acid. These interactions have been exploited to develop packaging, scaffolds, wound-healing, and dental implant materials with enhanced properties, and to remove and degrade water pollutants, among others.

Table 3. Some immunological proteins–TiO_2 interaction.

Immunological Protein	Abbreviation	Binding Energy	Intermolecular Energy
Intercellular adhesion molecule 1	ICAM−I	−11.63	−12.73
Mitogen−activated protein kinases	P−38	−11.73	−12.83
The nuclear factor−kB	NF−kB	−8.29	−9.39
Cyclooxygenase 2	COX−2	NR	NR
Interleukin 8	IL−8	−4.04	−5.14
Placental growth factor	PIGF	−9.36	−10.36
C–X–C motif chemokine ligand 1	CXCL−I	1.67	0.57

Table 3. *Cont.*

Immunological Protein	Abbreviation	Binding Energy	Intermolecular Energy
C–X–C motif chemokine ligand 3	CXCL−3	NR	NR
C–X–C motif chemokine ligand 5	CXCL−5	576.34	575.34
C–X–C motif chemokine ligand 20	CCL−20	−8.25	−9.34
The cluster of differentiation 35	CD 35	5420	5420
The cluster of differentiation 66b	CD 66b	NR	NR
Matrix metallopeptidase 9	MMP−9	−9.01	−10.11

Adapted from Ranjan et al. [59]. NR: No reported.

4. Preparation of Functionalized Protein–TiO_2 Materials

The functionalization of protein-based materials through the introduction of organic (ascorbic acid, cellulose, and starch) and inorganic (metallic or metal oxide) compounds is an attractive way to fabricate protein-based hybrid materials with enhanced properties, which has seen a significant increase in the last few years [6]. The most common methods for developing functionalized protein-based materials are evaporative casting, dip-coating, layer-by-layer assembly, freeze-drying, electrospinning, and electrochemical through protein denaturation by gelation-coagulation process [6,45].

4.1. Evaporative Casting Method

The evaporative casting method is generally accepted and commercially used for its simplicity, flexibility, and applicability to large-scale production. It consists of preparing a viscous solution by mixing the components, casting them in a plate, and evaporating them under controlled temperature and vacuum conditions to remove the solvent solution and form film and coatings (Figure 1). In general, it is a relatively low-cost method (one-third to half of the other methods); however, its main limitations are the difficulty in achieving a uniform distribution of the reinforcement agent, the presence of air bubbles, and possible reactions between the polymeric matrix and functional agent [60].

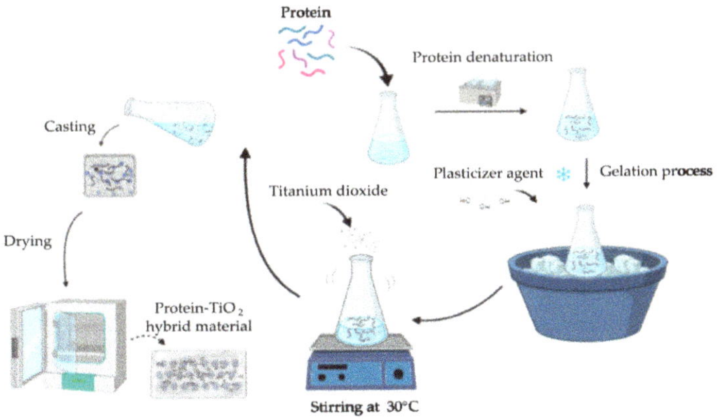

Figure 1. Schematic representation to laboratory scale of an evaporative casting method to prepare protein-based hybrid materials (adapted from Fan et al. [27], Al-Zoubi et al. [6]) (figure created with BioRender.com).

4.2. Dip Coating Method

Dip-coating is a technique widely used in many industrial fields to deposit onto any substrate. The process could be defined as depositing aqueous-based liquid phase coating solutions onto the surface of any substrate and is divided into five stages: immersion, start-up, deposition, drainage, and evaporation. It is achieved at low processing temperatures and is a low-cost method to develop thin coatings with high purity, good adhesion, high surface, and uniformity. However, this methodology requires high sintering temperatures and thermal expansion mismatch [61–63].

4.3. *Layer-by-Layer Deposition Method*

The layer-by-layer deposition is a common method for coating substrates to develop functional thin films. It is a cyclical process in which a charged material is adsorbed onto a substrate, and after washing, an oppositely charged material is adsorbed on the surface of the first layer. This constitutes a single bilayer film with a thickness generally on the order of nanometers, and the deposition process can be repeated until a multilayer film is obtained. This method offers advanced composites with exceptional properties (mechanical, electrical, optical, and biological) unavailable by other means, but this deposition process is complex, and the need for multiple dipping cycles hampers its usage in microtechnologies and electronics [64,65].

4.4. *Freeze-Drying Method*

Freeze-drying is a process that consists of removing the solvent from a frozen suspension containing mixed components. First, the gels are frozen, transforming the gel to a solid; then, sublimation of the solvent (mainly water) is then achieved at low pressure, avoiding the formation of the vapor–liquid interface. This method is widely used for aerogel preparation with highly porous and large specific surface area structures that allow rapid disintegration. However, this procedure requires sophisticated equipment compared to the evaporative casting method [66,67].

4.5. *Electospinning Method*

Electrospinning is a simple method to produce ultra-thin fibers with high surface area, highly porous structure, and small pore size. In this method, the mixed solution is pumped through a capillary conductive needle to form a droplet; under suitable conditions, solvent evaporation occurs, and the compound contracts into solid polymeric materials instead of fibers. It has the advantages of mild experimental conditions, low cost, easy operation and function, and a wide range of raw materials. The spinning process is controllable, and the parameters can be adjusted according to the different requirements in various research fields. However, electrospinning with raw materials that have a low molecular weight is difficult [68].

4.6. *Electrochemical Method*

Electrochemical methods are widely used for the preparation of thin films and coatings through anodic or cathodic techniques. Both processes are commonly used to prepare coatings by electrodeposition which include: electrophoretic process (EPD) using deposition of charged particles in a stable colloidal suspension on a conductive substrate, acting as one of the two oppositely charged electrodes in the EPD cell, and the electrolytic process (ELD), which starts from solutions of metal salts. They exhibit some advantages like low-cost, ability to coat complex shapes, speed, uniform coating thickness, rapid deposition rates, and the ability to coat complex substrates; however, it is difficult to produce crack-free coatings, it requires high sintering temperatures, and the bonding strength between coating and substrate is not strong enough [61,69].

5. Applications of Protein–TiO_2 Hybrid Composites

Protein-based materials exhibited a wide range of applications. However, most of their potential uses are limited by their poor physicochemical properties [35]. Thus, their functionalization with TiO_2 is a viable alternative to improve the technological and functional properties of protein-based materials such as gelatin, wheat gluten, kefiran, zein, and soy and whey protein isolates for several applications [49] (Figure 2), as discussed below.

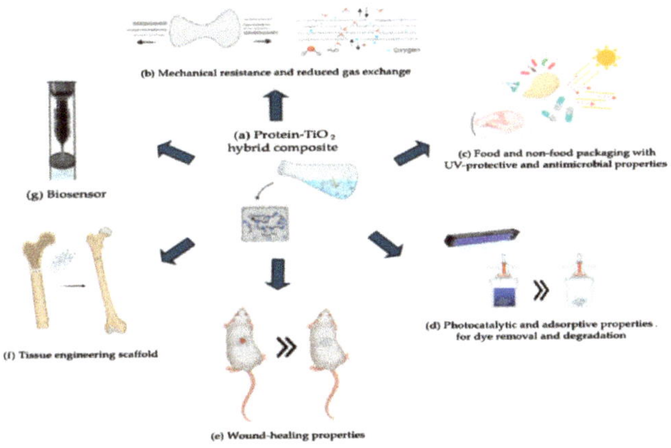

Figure 2. Protein–TiO_2 hybrid composites (**a**) with enhanced mechanical and reduced gas exchange (**b**) and their applications: as food and non-food packaging with UV-protective and antimicrobial properties (**c**), photocatalytic activity for dye removal and degradation (**d**), wound-healing material (**e**), tissue engineering scaffolds (**f**), and for the development of biosensors (**g**) (adapted from Lin et al. [3], Fan et al. [27], Alizadeh-Sani et al. [28], Emregul et al. [70], Ferreira et al. [71]) (figure created with BioRender.com).

5.1. Gelatin–TiO_2 Hybrid Composite

In the last years, the number of applications of gelatin-based materials has considerably increased. Gelatin is a protein obtained from the hydrolysis of collagen from mammalian sources, mainly pork and cattle. It is non-toxic, biodegradable, and biocompatible [72]. However, its main disadvantage for industrial applications (e.g., food packaging) is its hydrophilicity [73]. Therefore, the incorporation of TiO_2 into the gelatin matrix is a viable strategy to improve its technological and functional properties [74]. The most common method for the preparation of gelatin–TiO_2 hybrid composites is evaporative casting for films and coatings and freeze-drying for aerogels. Furthermore, the nanoparticles used are commercially available with sizes ranging from 10 to 25 nm in its anatase phase, in some cases in its rutile phase, using concentrations ≤1% in weight of total solid content, as shown in Table 4.

Table 4. Effect of TiO_2 incorporation on gelatin matrix properties.

Application	Method/Presentation	* Composition	TiO_2 Specifications	Relevant Results	Ref.
Food and non-food packaging	Evaporative casting/Film	Gelatin (4 g 100 mL^{-1}), glycerol (30% *w/w*)	Commercial SM: Hydrothermal (TiO_2): 0.5% *w/w* Size: 25 nm CP: Anatase	TiO_2 enhanced the physicochemical and antimicrobial properties of gelatin film.	[1]
Food and non-food packaging	Evaporative casting/Film	Fish gelatin (2.3% *w/v*), chitosan (1% *w/v*), glycerol (1% *w/v*)	(TiO_2:Ag): 0.4% *w/w*	Hybrid films showed improved physicochemical and antimicrobial properties.	[3]
Food and non-food packaging	Evaporative casting/Film	Gelatin (4 g 100 mL^{-1}), glycerol (25% *w/w*)	Commercial (TiO_2): 1% *w/w* Size: <10 nm CP: Anatase	TiO_2 improved UV-barrier, thermal, mechanical, and water-related properties of gelatin film.	[12]
Food and non-food packaging	Evaporative casting/Film	CMC (1 g 100 mL^{-1}), gelatin (1 g 100 mL^{-1})	Commercial (TiO_2:Ag): 0.4% *w/w* Size: 20 nm	TiO_2 improved the technological and photocatalytic properties of gelatin film.	[18]

Table 4. *Cont.*

Application	Method/Presentation	* Composition	TiO_2 Specifications	Relevant Results	Ref.
Food and non-food packaging	Evaporative casting/Film	Gelatin (4 g 100 mL^{-1}), glycerol (15% *w*/*w*)	SM: Sol–gel (TiO_2:Ag): 1% *w*/*w* Size: 10–20 nm CP: Anatase Body-centered tetragonal crystal structure	TiO_2 improved the technological properties of the gelatin film.	[20]
Food and non-food packaging	Evaporative casting/Film	Agar (1.5 g 100 mL^{-1}), gelatin (4 g 100 mL^{-1}), glycerol (35% *w*/*v*)	Commercial (TiO_2): 0.5% *w*/*w* Size: 10–20 nm CP: Anatase-Rutile	The hybrid film showed a marked UV-protective effect and improved water resistance.	[21]
Food and non-food packaging	Evaporative casting/Film	Gelatin (15 mg·mL^{-1})	(TiO_2): 0.5% *w*/*w* Size: 12.2 nm CP: Anatase Crystal structure	The film exhibits antibacterial activity.	[26]
Food and non-food packaging	Evaporative casting/Film	Gelatin (3 g 80 mL^{-1}), PVA (3 g 80 mL^{-1}), glycerol (30% *w*/*w*)	Commercial (TiO_2:4A zeolite): 1% *w*/*w*	Functionalization improved the physicochemical and antimicrobial properties of the gelatin–PVA film.	[74]
Food and non-food packaging	Evaporative casting/Coating	Gelatin (3 g 80 mL^{-1}), PVA (3 g 80 mL^{-1}), glycerol (30% *w*/*w*)	Commercial (TiO_2:4A zeolite): 1% *w*/*w*	The hybrid film effectively extended the shelf life of white shrimp.	[75]
Food and non-food packaging	Evaporative casting/Film	Gelatin (8% *w*/*w*), sorbitol: glycerol ratio 3:1 (40% *w*/*w*)	Commercial (TiO_2): 1% *w*/*w* Size: 10–15 nm CP: Anatase-Rutile	The hybrid film showed antimicrobial properties.	[76]
Food and non-food packaging	Evaporative casting/Film	Gelatin (1 g 100 mL^{-1}), CMC (1 g 100 mL^{-1})	Commercial (TiO_2:Ag): 0.4% *w*/*w* Size: 21 nm CP: Anatase	Hybrid films showed improved physicochemical and antimicrobial properties.	[77]
Food and non-food packaging	Evaporative casting/Film	Gelatin (4% *w*/*w*), agar (1.5% *w*/*v*), glycerol (35% *w*/*w*)	Commercial (TiO_2): 2% *w*/*w*	Gelatin–TiO_2 effectively delayed fish oil oxidation.	[78]
Biomedical	Freeze-drying/Hydrogel	Sodium alginate (2% *w*/*v*), gelatin (0.5% *w*/*v*), β-tP (1% *w*/*v*)	SM: Electrochemical anodization (TiO_2): 0.1% *w*/*v* Size: 110 nm CP: Anatase Nanotubes	Hybrid hydrogel had adequate porosity and mechanical resistance.	[79]
Biomedical	NI/Scaffold	NI	SM: Biometic (TiO_2): NI Size: 30–35 nm CP: Anatase	Hybrid scaffold promoted osteointegration and enhanced bone regeneration.	[71]
Biomedical	NI/NI	Gelatin (2 mg·mL^{-1})	Electrochemical anodization (TiO_2): NI Size: 100 nm CP: Rutile Nanotubes (20 nm × 350 nm) Low crystal structure	Hybrid material could potentially be used for orthopedic and dental applications.	[80]
Biomedical	Electrochemical deposition/coating	Hap (NI), Gelatin (100 mg 100 mL^{-1}), GO (2 mg mL^{-1})	Hydrothermal (TiO_2): NI Crystal structure	The hybrid coating showed excellent biocompatibility with MC3T3-E1 cells.	[81]
Biomedical	Freeze-drying/Hydrogel	Gelatin (2 g 100 mL^{-1})	(TiO_2): 0.5% *w*/*v*	The hybrid composite had better wound-healing properties than gelatin film.	[73]

Table 4. *Cont.*

Application	Method/Presentation	* Composition	TiO_2 Specifications	Relevant Results	Ref.
Biomedical	Polymer blend/Biosensor	CMC:Gelatin (3.75 mg), solution of superoxide dismutase (4733 U, 1 mg), glutaraldehyde (0.005 M)	SM: Hydrothermal (TiO_2): 0.1% *w/w* Size: 50 nm CP: Anatase	The biosensor exhibited high analytical performance, high sensitivity, and fast response time for superoxide radical detection.	[70]
Pharmaceutical	NI/Capsule	Gelatin (NI)	Commercial (TiO_2): 3.5% *w/w* Size: 177.2 nm CP: Anatase Crystalline structure	The capsules could be printed gray by UV-laser.	[82]
Metal corrosion resistance	NI/Coating	Gelatin (8 wt.% in 20 wt.% acetic acid)	Commercial (TiO_2): 3% *w/w* Size: 10–25 nm CP: Anatase Purity: >99% Density: 3.9 $g \cdot cm^{-3}$	Gelatin–TiO_2 composite improved the corrosion resistance of steel material.	[83]
Hydrogen production	NI/microspheres	Gelatin (5 g 100 mL^{-1})	SM: Sol–gel Titania precursor (10 mL of tetra-n-butyl titanate in 50 mL of ethyl alcohol) Size: 50–100 nm CP: Anatase High crystallinity and purity	Gelatin improved the adsorptive properties of TiO_2.	[84]

* Material composition was based on the best-reported results. NI: No information; CMC: carboxymethyl cellulose; PVA: polyvinyl alcohol; GO: graphene oxide; β-tP: β-tricalcium phosphate; Hap: hydroxyapatite; SM: synthesis method; (TiO_2): concentration of titanium dioxide; CP: crystallite phase.

5.1.1. Food and Non-Food Packaging Applications of Gelatin–TiO_2 Hybrid Composite

The potential use of gelatin-based materials functionalized with TiO_2 nanoparticles as food and non-food packaging material has been extensively explored [26,76]. Nassiri and Nafchi [76] developed a bovine gelatin film reinforced with TiO_2 nanoparticles with antimicrobial properties against *S. aureus* and *E. coli*, associated with the physical and chemical interactions of TiO_2 with the bacteria cell membrane. Incorporation of TiO_2 at low concentrations (5% *w/w*) decreases the water vapor (from 8.90 to 1.61×10^{11} $g \cdot m^{-1} \cdot s^{-1} \cdot Pa^{-1}$), and oxygen permeability (from 214 to 95 $cm^3 \cdot \mu m/m^2 \cdot day$) of protein-based film. Similarly, Qingyan et al. [26] informed that gelatin–TiO_2 film exhibited antimicrobial activity against *E. coli* (54% inhibition of viable cells) and *S. aureus* (44% inhibition of viable cells) under UV-light irradiation (365 nm) after 120 min of exposure. The above, associated with the photocatalytic properties of TiO_2 and its ability to generate reactive molecules (hydrogen peroxide, hydroxyl radical, and superoxide anions) with antimicrobial properties by affecting the cell viability. Moreover, the addition of TiO_2 (1% *w/w* of total solid content) in the gelatin film promoted an increase in its mechanical and thermal properties. It decreased water solubility, moisture uptake, water vapor permeability, and transparency due to the formation of hydrogen and Ti–O–C bonds and electrostatic interactions between protein and inorganic nanoparticles [12].

Azizi-Lalabadi et al. [74] made a hybrid film composed of gelatin and polyvinyl alcohol, reinforced with TiO_2 nanoparticles previously embedded in 4A-zeolite. The enhanced physicochemical (optical, gas-barrier, and water resistance) were attributed to the interaction of the N–H functional group present in the protein structure, with TiO_2 through hydrogen bonds. Moreover, the hybrid film exhibited antimicrobial properties especially against Gram-negative bacteria (*E. coli* and *P. fluorescens*). Moreover, the hybrid film effectively extended the shelf life of white shrimp (up to 12 days) compared to uncoated samples (6 days), without significant changes in sensory attributes [75]. Likewise, Riahi et al. [1] fabricated an active gelatin–TiO_2–grape seed extract film for food packaging purposes and found that water contact angle, water vapor permeability, mechanical properties, and UV-protective effect improved in a dose-dependent response with an optimum TiO_2 concentration of 0.5% *w*/*w*, which was attributed to the chemical interaction of TiO_2 and C=O groups in the protein structure. On the other hand, the hybrid film exhibited antimicrobial activity in strain- and TiO_2 dose-dependence, where the Gram-negative bacteria were less susceptible than Gram-positive. At low concentrations of TiO_2 (<3% *w*/*w*), the hybrid film showed a bacteriostatic effect against *E. coli* and *L. monocytogenes*, while at 5% *w*/*w* exhibited bactericidal action.

Pirsa et al. [77] evaluated the antioxidant and antimicrobial properties of a carboxymethyl cellulose–gelatin film reinforced with TiO_2:Ag-doped nanoparticles. The hybrid film exhibited better mechanical properties (greater flexibility) in comparison with the control group. Moreover, it showed antioxidant activity and antibacterial effect against *E. coli* and *S. aureus* in a TiO_2:Ag concentration-dependent response. Furthermore, the carboxymethyl cellulose (CMC)–gelatin–TiO_2: Ag exhibited good photocatalytic degradation of ethanol, benzene, and ammonia [18]. Furthermore, the incorporation of TiO_2: Ag-doped nanoparticles improved the antioxidant, mechanical, UV-barrier, water resistance, and mechanical properties of a *Rhinobatos cemiculus* gelatin film in a dose-dependent manner. At a low concentration of TiO_2, it can disperse uniformly and insert in the amorphous region of soy protein isolate (SPI), leading to a major interaction between both components; however, at high concentrations of TiO_2, it could cause agglomerations interfering with the organization and interaction of protein and TiO_2 [16].

Similar results were reported in a fish gelatin–chitosan film functionalized with TiO_2:Ag nanoparticles, where the improved antibacterial activity (*E. coli*, *S. aureus*, and *Botrytis cinerea*), optical, water-related, and mechanical properties were in a TiO_2:Ag dose-dependent response [3]. The addition of TiO_2:Ag-doped nanoparticles did not alter the typical structure of biopolymers, but instead promoted stronger intramolecular hydrogen bonds formation [16,85]. On the other hand, it has been reported that the improved UV-protective effects, water-related, and mechanical properties of a fish gelatin–agar–TiO_2 film could be negatively affected by a high concentration of TiO_2 (>0.5 g 100 mL^{-1}), mainly by an inhomogeneous dispersion and saturation of nanoparticles in the protein structure [21].

Additionally, Vejdan et al. [78] informed that a gelatin–agar bilayer film functionalized with TiO_2 nanoparticles effectively delays fish oil photo- and auto-oxidation up to 18 days. They reported that hybrid film containing 2% of TiO_2 could control fish oil oxidation due to the enhanced UV-protective and oxygen-barrier properties associated with the physicochemical characteristics of TiO_2.

According to the results, incorporation of TiO_2 into the gelatin-matrix improved its mechanical, thermal, UV-protective, gas-barrier, and water-related properties with antioxidant and antimicrobial performance, both desirable characteristics for the development of food and non-food packaging materials.

5.1.2. Biomedical Applications of Gelatin–TiO_2 Hybrid Composite

Gelatin–TiO_2 hybrid composites have been used for biomedical purposes. Lai et al. [80] immobilized gelatin onto TiO_2 nanotubes to modulate osteoblast behavior for orthopedic and dental applications. The authors found that cell spreading, proliferation, and differentiation of osteoblasts were improved by gelatin–TiO_2 hybrid material. They argued that extracellular matrix protein-based plays an important role in bone mineralization, while TiO_2 present in the hybrid matrix facilitates osteoblast differentiation. Ferreira et al. [71] fabricated a macroporous TiO_2–functional hydroxyapatite–gelatin scaffold loaded with multipotent adult progenitor cells for bone regeneration applications in calvaria defects. They informed that a hybrid scaffold promoted osteointegration and enhanced bone regeneration with complete closure defect. The result was associated with the ability of TiO_2 to form complexes with calcium ions, promoting the adsorption of calcium-binding extracellular matrix proteins and Argine-Glycine-Aspartate specific peptide sequences.

Additionally, hydroxyapatite–gelatin–graphene oxide composite deposited on TiO_2 nanotubes by electrochemical deposition exhibited excellent biocompatibility with MC3T3-E1 cells, promoting a better cellular integration [81]. Moreover, Urruela-Barrios et al. [79] mentioned that a sodium alginate–gelatin hydrogel 3D printing functionalized with nano-TiO_2 and β-tricalcium phosphate exhibited a potential use for tissue engineering application. The hybrid material fabricated with the micro-extrusion process, exhibited adequate porosity (pore size ranged from 150 to 240 μm), and mechanical resistance (13 MPa) to promote cell proliferation and cartilages.

Nikpasand and Reza-Parvizi [73] evaluated in vivo the wound dressing properties of a gelatin–TiO_2 hybrid hydrogel in an open and infected with *S. aureus* methicillin-resistant at 5×10^7 colony forming units (CFU) by excision-type wound-healing study in rats. They found that the hybrid composite exhibited a good wound-healing effect (wound area closure of 100% after 21 days), in comparison with gelatin-wound treatments (wound area closure of 71% after 21 days). Nonetheless, animals treated with the hybrid composite did not show wound infection by pathogenic bacteria after 14 days of evaluation and exhibited accelerated re-epithelization through fibroblast proliferation without inflammatory response after 21 days, which could be considered for wound therapies. On the other hand, Emregul et al. [70] developed a carboxymethyl cellulose–gelatin–TiO_2–superoxide dismutase biosensor supported in Pt surface for $O_2^{\bullet-}$ detection. They reported that the biopolymer blend (CMC and gelatin), provided a biocompatible environment for super oxide dismutase–TiO_2, which acts as a nanoscale electrode, enhancing the electron transfer rate through the Pt electrode. The hybrid sensor exhibited high analytical performance with a wide linear range of 1.5 nM to 2 mM, and high sensitivity and fast response time (1.8 s) for $O_2^{\bullet-}$ detection in healthy and cancerous brain tissue (coefficient of determination or R^2 of 0.991). In this context, functionalization of gelatin-based materials with TiO_2 exhibited potential biomedical applications, associated with its enhanced biological properties.

5.1.3. Other Applications of Gelatin–TiO_2 Hybrid Composite

Other investigated applications of the gelatin–TiO_2 hybrid composite include pharmaceutical (development of empty capsule shells), anti-corrosive material, and hydrogen storage. Hosokawa et al. [82] evaluated the application of UV-laser irradiation (at 355 nm) to print hard gelatin capsule shells with TiO_2, and it was found that hybrid capsules could be printed gray in a laser power-dependent response.

Additionally, Hayajneh et al. [83] studied the effect of gelatin–TiO_2 hybrid coating on the corrosion resistance of AISI 304 stainless steel, in a simulated marine environment (solution with NaCl at 3.5% *w*/*v*) through potentiodynamic polarization studies. The presence of hybrid coating improved the corrosion resistance of steel material (corrosion rate 2.63×10^{-3} mpy) in comparison with gelatin-coated (corrosion rate 10.10×10^{-3} mpy) and uncoated (corrosion rate 9.94×10^{-3} mpy) steel. The results were associated with the formation of a dense and stable network structure formed by the gelatin and TiO_2 nanoparticles.

Furthermore, Bin Liu et al. [84] used gelatin as a template to fabricate TiO_2 mesoporous microspheres for hydrogen production. They reported that the assistance of gelatin positively influenced the morphology and physicochemical characteristics of TiO_2 nanoparticles (surface area of $98.3\ m^2 \cdot g^{-1}$ and pore size of 11.9 nm), enhancing the hydrogen adsorption capacity and hydrogen storage performance of hybrid microspheres. However, its hydrogen adsorption mechanism remains unclear. According to these data, the gelatin–TiO_2 hybrid material exhibited pharmaceutical, anti-corrosive, and hydrogen production applications.

5.2. Whey Protein–TiO_2 Hybrid Composite

Whey protein is a by-product obtained from dairy processing during cheese production. It is used to develop edible films and coatings with good biodegradability and lower gas permeability for diverse applications [29]. However, the potential uses of whey protein-based materials are limited by their higher hydrophilicity due to polar residues outside the globular structure, which causes softening when they come in contact with high-moisture environments [86]. On the other hand, it exhibited good biocompatibility to interact with inorganic compounds like TiO_2 to improve its technological and functional properties [29]. The most common method for preparing whey protein–TiO_2 hybrid composites is evaporative casting. Furthermore, the nanoparticles used are commercially available with sizes ranging from 10 to 25 nm in its anatase phase and using concentrations ≤1% in weight of total solid content, as listed in Table 5.

Table 5. Effect of TiO_2 incorporation on whey protein matrix properties.

Application	Method/Presentation	* Composition	TiO_2 Specifications	Relevant Results	Ref.
Food and non-food packaging	Evaporative casting/Film	WPI (10% *w/w*)	Commercial (TiO_2): 0.25% *w/w* Size: 50–100 nm CP: Anatase Purity: >98.5%	TiO_2 improved the physicochemical properties of whey protein film.	[17]
Food and non-food packaging	Evaporative casting/Film	WPI (10% *w/v*), cellulose (1% *w/v*), glycerol (6% *w/v*), REO (2% *w/v*)	Commercial (TiO_2): 1% *w/v* Size: 10–25 nm CP: Anatase	Coated meat exhibited microbial stability during cold storage.	[28]
Food and non-food packaging	Evaporative casting/Film	WPI nanofibers (5% *w/v*), glycerol (4% *w/v*),	Commercial (TiO_2): 1% *w/w* Size: 20 nm Nanotubes Purity: >99%	The hybrid film effectively extends the shelf life of chilled meat.	[29]
Food and non-food packaging	Evaporative casting/Film	WPI (5% *w/v*), kefiran (5% *w/v*), glycerol (35% *w/w*)	Commercial (MMT-TiO_2): 1% *w/w* CP: Anatase	TiO_2 improved the physicochemical properties of kefiran–whey protein film.	[31]
Food and non-food packaging	Evaporative casting/Film	WPI (10% *w/v*), cellulose (1% *w/v*), glycerol (6% *w/v*), REO (2% *w/v*)	Commercial (TiO_2): 1% *w/v* Size: 10–25 nm CP: Anatase Purity: >99%	The hybrid film exhibited antimicrobial and antioxidant properties.	[86]
Food and non-food packaging	Evaporative casting/Film	WPI (5% *w/v*), TiO_2 (1% *w/w*), glycerol (5% *w/v*)	Commercial (TiO_2): 1% *w/w* Size: <20 nm CP: Anatase	TiO_2 improved the physicochemical properties of whey protein film.	[87]
Food and non-food packaging	Evaporative casting/Film	WPI (10% *w/v*), cellulose (1% *w/v*), glycerol (6% *w/v*), REO (2% *w/v*)	Commercial (TiO_2): 1% *w/v* Size: 10–25 nm CP: Anatase Purity: >99%	Meat treated with the hybrid film showed reduced lipid peroxidation during cold storage.	[88]
Food and non-food packaging	Evaporative casting/Film	Chitosan (1.5 g 50 mL^{-1} of acetic acid), WPI (0.5 g 50 mL^{-1} of water)	Commercial (TiO_2): 0.01 g CP: Anatase Crystalline structure	The hybrid film exhibited improved physicochemical properties.	[89]
Food and non-food packaging	Evaporative casting/Film	WPI (5% *w/v*), kefiran (5% *w/v*), glycerol (35% *w/w*)	Commercial (MMT-TiO_2): 1% *w/w* Size: 20 nm CP: Anatase	The hybrid film exhibited improved physicochemical properties.	[90]
Food and non-food packaging	Evaporative casting/Film	WPI (3% *w/v*), chitosan (10 g/L), ZMEO (1% *v/v*), glycerol (30% *w/w*)	Commercial (TiO_2): 2% *w/w* CP: Anatase-Rutile	The hybrid film exhibited antimicrobial activity.	[91]
Textile	Dip-pad-dry-cure process/Coating	WPI (3% *w/v*), cotton fabrics (200 g/m^2)	Commercial (TiO_2): 6% *w/w*	The hybrid coating exhibited improved antimicrobial activity.	[92]

* Material composition was based on the best-reported results. NI: No information; WPI: whey protein isolate; REO: rosemary essential oil; MMT: montmorillonite; ZMEO: *Zataria multiflora* essential oil.; SM: synthesis method; (TiO_2): concentration of titanium dioxide; CP: crystallite phase.

5.2.1. Food and Non-Food Packaging Applications of Whey Protein–TiO_2 Hybrid Composite

The potential use of whey protein–TiO_2 hybrid material for food packaging purposes has been investigated [28], as shown in Table 5. Zhou et al. [87] prepared a biodegradable whey protein film functionalized with TiO_2. It was found that technological properties such as UV-protective, mechanical, and water-resistance properties were improved in a TiO_2 dose-dependent response, associated with the intramolecular connections of protein and TiO_2 through covalent and non-covalent interactions. Moreover, the authors argued that at low concentrations of TiO_2, a reinforcement of whey protein–TiO_2 structure occurs. Meanwhile, self-assembly of TiO_2–TiO_2 interactions are detected at high TiO_2 concentrations, influencing its technological and functional properties, mainly associated with a reduction in the crystalline structure of TiO_2 by its incorporation in a polymeric matrix and its tendency to form agglomerates at higher concentrations [17,31]. Similar trends were informed in a kefiran–whey protein film functionalized with TiO_2, where an excessive amount of TiO_2 in the polymeric matrix affected its functionality because TiO_2 may act as an anti-plasticizer agent [31,90]. Moreover, in a combined chitosan–whey protein film reinforced with sodium laurate–TiO_2 nanoparticles. Zhang et al. [89] reported that sodium laurate-modified TiO_2 incorporation influenced the transparency, water vapor permeability, and mechanical and thermal properties of the hybrid film in a dose-dependent manner, and its intermolecular interaction with the available functional groups of the chitosan–whey protein matrix. Gohargani et al. [91] fabricated a chitosan–whey protein film, functionalized with TiO_2 and *Zataria multiflora* essential oil (ZMEO) nanoparticles with enhanced antimicrobial properties against foodborne pathogenic bacteria such as *L. monocytogenes*, *S. aureus*, and *E. coli*. Results were attributed to the synergistic effect of bioactive compounds present in the ZMEO and TiO_2 nanoparticles. Moreover, the TiO_2–ZMEO incorporation into the hybrid film, improved water vapor permeability, and tensile strength with a significant decrease in the film's transparency and color, associated with the physicochemical properties of TiO_2.

Alizadeh-Sani et al. [28] informed that a whey protein isolate–cellulose nanofiber-TiO_2–rosemary essential oil (REO) effectively preserved quality (microbial deterioration and sensory attributes) of refrigerated meat during cold storage. They reported that lamb meat treated with the hybrid film showed microbial stability (4.1 log·CFU·g^{-1} of viable cells) for 6 days at 4 °C storage without changes in sensory attributes (color, odor, texture, and overall acceptability). Moreover, the treated meat exhibited reduced lipid oxidation during storage, ascribed to antioxidant properties of REO (80% of radical scavenging) [88]. Furthermore, the TiO_2 (1% *w*/*w*) and REO (2% *w*/*w*) addition in the whey protein isolate/cellulose nanofiber hybrid film, improved mechanical (tensile strength, elongation at break, and elastic modulus) and water-related properties (moisture uptake, water solubility, and water vapor permeability), with a decrease in its transparency in a dose-dependent response in comparison with whey protein-based film, associated with the UV-scattering ability of TiO_2. Furthermore, the hybrid film showed an antimicrobial effect against foodborne bacteria (*E. coli* O157:H7, *L. monocytogenes*, *P. fluorescens*, and *S. enteritidis*) in a strain-dependent manner. It was associated with antimicrobial properties of TiO_2 and bioactive compounds (1,8-cineole, α-pinene, and camphor) in the REO; which can alter the cell membrane and finally cause cell death [86]. Nonetheless, they informed that a low content of TiO_2 migrated from the polymeric matrix to the meat product, under the Food and Drug Administration limit recommendations (<1% *w*/*w*) [88]. Similarly, Feng et al. [29] informed that a whey protein–TiO_2 hybrid film is effective in extending the shelf life of chilled meat (up to 15 days) without significant changes in its quality parameters (weight loss less than 7.87%, reduced lipid peroxidation, and microbial stability) during cold storage (4 °C). Moreover, the hybrid film exhibited enhanced mechanical, optical, and water-related properties associated with the physical and chemical interactions between carboxylic and sulfhydryl groups of some amino acids present in the protein matrix with TiO_2.

According to the evidence, the incorporation of TiO_2 into whey protein-based materials can improve the thermal, UV-barrier, mechanical, and water-related properties through physical and chemical interactions. Furthermore, whey protein films functionalized with TiO_2 exhibited antimicrobial properties for potential food and non-food packaging.

5.2.2. Other Applications of Whey Protein–TiO_2 Hybrid Composite

Ortelli et al. [92] fabricated a hybrid cotton fabric with anti-fire properties incorporating a whey protein–TiO_2 coating by the dip-pad-dry-cure process (Table 5). In general, the hybrid cotton material showed major durability (resistance to washing) and flame-resistant compared with the control group because TiO_2 acts as a physical reinforcement agent to fix whey protein to cotton fabrics in a stable way with the hydroxyl groups.

5.3. Collagen–TiO_2 Hybrid Composite

Collagen is a large, coherent, covalently crosslinked fibrillar network protein. Its main sources are porcine, bovine, and ovine with many applications in the food, cosmetics, pharmaceutical, and biomedical industries [34]. The disadvantages of collagen are poor thermal instability, poor mechanical properties, and the possible contamination by pathogenic bacteria and chemical substances [93]. Particularly, collagen has been combined with TiO_2 to improve its physicochemical properties [5]. Preparation of collagen–TiO_2 hybrid composites is usually by dip-coating, followed by freeze-drying for aerogel development. Furthermore, the nanoparticles used are commercially available or synthesized by the Sol–gel method with sizes ranging from 10 to 30 nm in its anatase phase, and in some cases in its rutile phase (Table 6).

Table 6. Effect of TiO_2 incorporation on collagen-based materials.

Application	Method/Presentation	* Composition	TiO_2 Specifications	Relevant Results	Ref.
Biomedical	Dip coating/Composite	Collagen-MWCNTs composite coated Ti incorporated with 20 μg/cm^2 of MWCNTs	Commercial (TiO_2): NI	The high roughness of hybrid materials improved cell proliferation.	[5]
Biomedical	Dip coating/NI	Volume ratio 1:1.5 GPTMS-TiO_2 solutions into a Collagen solution (3 mg·mL^{-1}) to cover the Mg alloys	SM: Sol–gel (TiO_2): NI TiO_2 with an amorphous structure	Protect alloy from corrosion, promote fibroblast proliferation.	[93]
Biomedical	NI/Film	Collagen (0.5 mg·m^{-1})	SM: Electrochemical deposition (TiO_2): NI TiO_2 with a crystalline structure	The hybrid film showed rapid cell adhesion and proliferation.	[94]
Biomedical	NI/Composite	Collagen (NI)	Commercial SM: Anodization (TiO_2): 0.3% *w/w* Size: 67 nm	Hybrid composite facilitated epithelial cell stretching and sheet formation.	[95]
Biomedical	Atomic layer deposition/Membrane	Collagen membrane (25 mm × 15 mm)	Commercial (TiO_2): NI	The hybrid membrane exhibited the proliferation of osteoblast.	[96]
Biomedical	NI/Composite	Mol ratio 1:1 of PdO–TiO_2 incorporated to g-PMMA–Collagen	SM: Sol–gel (TiO_2): NI Size: 8 nm CP: Anatase	TiO_2 incorporation improved thermal stability, mechanical strength, and enhancement of collagen.	[97]
Biomedical	Freeze-drying process/Aerogel	Collagen–PVP–TiO_2 1:20:0.5 mass ratio	SM: Sol–gel (TiO_2): NI Size: 24.4 nm CP: Anatase–Rutile	PVP improves the thermal stability and coercivity of the nanocomposite scaffold.	[34]
Biomedical	Freeze-drying process	Collagen–chitosan–TiO_2 1:1:0.1 mass ratio	SM: Sol–gel (TiO_2): NI Size: 20–30 nm CP: Anatase	TiO_2 improves mechanical properties, resistance to degradation, and antibacterial ability, and wound repair.	[27]
Non-food packing	NI/NI	Collagen (4 g 100 mL^{-1})	SM: Sol–gel (TiO_2): 2% *w/w* Size: 30 nm CP: Anatase	TiO_2 increases the thermal stability of collagen film improves and reduces UV light penetration, and solubility.	[98]

Table 6. *Cont.*

Application	Method/Presentation	* Composition	TiO_2 Specifications	Relevant Results	Ref.
Environmental remediation	Dip coating/NI	Collagen (template)	(TiO_2: Tb^{3+}): 2% *w*/*w* Size: 9.6 nm CP: Anatase	Collagen structure was preserved and photocatalytic performance of TiO_2 increased.	[99]
Electrochemical studies	Chemical reactions/NI	NI	SM: Template (TiO_2): NI Size: 10–20 nm CP: Anatase	Hybrid material showed excellent electrochemical lithium and sodium storage properties.	[100]

* Material composition was based on the best-reported results. NI: No information; MWCNTs: multiwalled carbon nanotubes; g-PMMA: poly(methylmethacrylate); GPTMS-TIP: (3-glycidoxypropyl)trimethoxysilane; PVP: poly(vinyl pyrrolidone); SM: synthesis method; (TiO_2): concentration of titanium dioxide; CP: crystallite phase.

5.3.1. Biomedical Applications of Collagen–TiO_2 Hybrid Composite

Table 6 lists, works on collagen-based materials functionalized with TiO_2 for biomedical applications. Park et al. [5] evaluated the effect of collagen-multi-walled carbon nanotubes (MWCNTs) composite coating deposited on titanium, using a dip-coating method on osteoblast growth. Cell proliferation studies confirmed a strong dependence of the extent of cell proliferation on the amount of MWCNTs incorporated in the composite in a dose-dependent response. Collagen–MWCNT–Ti showed higher cell proliferation than the collagen–MWCNT composite, where TiO_2 was responsible for cell proliferation. Truc et al. [94] studied the interaction between fibroblast and collagen modified on titanium (Ti) surface by electrochemical deposition (ECD), to reduce dental implant failure. They found that the Ti/Collagen hybrid composite showed rapid cell adhesion and proliferation.

Nojiri et al. [95] evaluated the establishment of perpendicularly oriented collagen attachments on TiO_2 nanotubes (TNT), which exhibited significant binding resistance, and the chemically linked collagen–TiO_2 facilitated epithelial cell stretching and sheet formation. Similarly, Bishal et al. [96] informed that collagen–TiO_2 promotes human osteoblast growth and proliferation in a dose-dependent manner with no inflammatory response detected, which was associated with the ability of TiO_2 to interact with calcium and phosphate elements, suggesting that this material could be used for applications in bone tissue engineering. On the other hand, Vedhanayagam et al. [97] informed that the poly(methyl methacrylate)–collagen–PdO–TiO_2 hybrid scaffolds did not show toxic effects on MG 63 cells (human osteosarcoma), and enhanced the alkaline phosphatase activity during in vitro osteogenic differentiation by the secretion of the osteogenic protein, leading to bone formation. Moreover, the hybrid scaffold exhibited higher thermal stability (83.45 °C), and mechanical strength (Young's modulus 105.57 MPa) than the pure collagen scaffold (71.64 °C, 11.67 MPa, respectively), due to the chemical and physical interaction between collagen and Palladium oxide (PdO)–TiO_2.

Additionally, collagen–silane–TiO_2 has also been used as a functional agent of Mg alloys. The hybrid composite promotes the formation of a stable $Mg(OH)_2/MgCO_3/CaCO_3$ structure that effectively protects its corrosion. Moreover, the collagen–silane–TiO_2 improved osteoblasts and fibroblasts proliferation compared to bare and silane–TiO_2-coated alloys. In the long term, collagen–silane–TiO_2 is a viable strategy to prevent Mg alloy degradation due to the formation of a complex structure [93]. On the other hand, Li et al. [34] made 3D nanocomposite scaffolds composed of collagen, polyvinyl pyrrolidone (PVP), and TiO_2 nanoparticles, with good degradation resistance in PVP dose-dependent response for potential tissue engineering applications. Likewise, collagen–chitosan–TiO_2 scaffolds exhibited antimicrobial activity against *S. aureus* and improved permeability, stability to degradation, and cell aggregation to stop bleeding, which are suitable for the development of wound-healing materials [27].

Significant evidence shows that collagen functionalization with TiO_2 nanoparticles improved its biological properties for dental implants and bone and dermal regeneration.

5.3.2. Other Applications of Collagen–TiO_2 Hybrid Composite

Other researched applications of the collagen–TiO_2 hybrid composite include the development of packaging materials, catalysts, and electronics (Table 6). Erciyes et al. [98] proposed the use of leather solid wastes as a source of collagen hydrolyzed to make composites functionalized with TiO_2. The hybrid film exhibited improved water vapor permeability, water-solubility, elongation at break, and tensile strength. The authors highlighted the potential reuse of collagen-waste to develop packaging materials.

Additionally, Luo et al. [99] informed that collagen–TiO_2: Tb^{3+}-doped hybrid material exhibited excellent photocatalytic performance against methyl orange (93.87%) dye after 6 h of exposure in UV-light irradiation (150 W).

Furthermore, Cheng et al. [100] proposed a facile synthetic strategy to engineer a one-dimensional (1D) hierarchically ordered mesoporous TiO_2 nanofiber bundles (TBs) by using low-cost natural collagen fibers as a bio-template. In general, the hybrid structure can offer shortened ion diffusion paths, ensuring an efficient electrolyte penetration for ion access without affecting its structural integrity. They conclude that the hybrid materials had excellent electrochemical lithium and sodium storage properties.

In general, the collagen–TiO_2 hybrid material exhibited potential applications such as food and non-food packaging, environmental remediation, and electrochemical studies.

5.4. *Soy Protein–TiO_2 Hybrid Composite*

Soy protein isolate (SPI) is a by-product attained from the manufacture of soybean oil with a complex mixture of proteins (β-conglycinin and glycinin) with a minimum protein content of 90% on a moisture-free basis [101,102]. It is readily available, biodegradable, and biocompatible for edible coatings [19] with potential usage on food packaging [72,103]. However, the main disadvantages of SPI-based films include weak mechanical properties and high sensitivity to humidity [102,104]. In that sense, SPI films have been functionalized with TiO_2 to enhance their physical properties, where the most common method for its preparation is evaporative casting. Furthermore, the nanoparticles used are commercially available in its anatase phase, with concentrations ranging from 0.5% to 2% in weight of total solid content (Table 7).

Table 7. Effect of TiO_2 incorporation on soy protein isolate-based materials.

Application	Method/Presentation	* Composition	TiO_2 Specifications	Relevant Results	Ref.
Food packaging	Evaporative casting/Films	Soy protein isolate (5 g 100 mL^{-1}) glycerol (0.4 g)	Commercial (TiO_2): 1.5% *w*/*v* CP: Anatase	TiO_2 improved the physicochemical and antimicrobial properties of the soy protein isolate film.	[101]
Food packaging	Evaporative casting/Films	Soy protein isolate (5 g 100 mL^{-1}), sorbitol (20%), glycerol (10%)	Commercial (TiO_2): 2% *w*/*w*	The hybrid film exhibited improved mechanical properties.	[102]
Food packaging	Evaporative casting/ Films	Soy protein isolate (5%), glycerol (2%),	Commercial (TiO_2): 0.5% *w*/*w* Size: 15–30 nm CP: Anatase	TiO_2 improved the physicochemical and antimicrobial properties of the soy protein isolate film.	[19]
Food packaging	Evaporative casting/Films	Soy protein isolate (4.5 g 150 mL^{-1}), glycerol (3.75 g 150 mL^{-1})	(TiO_2): 1.33% *w*/*w*	Hybrid composite effectively extended the shelf life of strawberries and antimicrobial activity.	[105]
Food packaging	Evaporative casting/ Films	Soy protein isolate (NI)	(TiO_2): NI	Grapes treated with hybrid films showed higher quality parameters than uncoated fruits.	[106]
Food and non-food packaging	Evaporative casting/Films	Soy protein isolate (4.5 g 150 mL^{-1}), glycerol (2%)	Commercial (TiO_2): 1.33% *w*/*w* TiO_2 with crystalline structure	The hybrid composite showed antimicrobial activity.	[23]

* Material composition was based on the best-reported results. NI: No information; SM: synthesis method; (TiO_2): concentration of titanium dioxide; CP: crystallite phase.

5.4.1. Food and Non-Food Packaging Applications of Collagen–TiO_2 Hybrid Composite

Table 7 lists the work on soy protein isolate–TiO_2 hybrid material for food and non-food packaging development with enhanced properties. Malathi et al. [102] informed that TiO_2 incorporation into an SPI film promotes an increase in thickness, opacity, tensile strength, and elongation at break of the cast film, which was associated with the hydrogen bonding or O–Ti–O bonding. Moreover, a strong charge and polar interaction between side chains of soy protein molecules restrict segment rotation and molecular mobility, leading to an increase in the elongation of the hybrid film. Furthermore, Lu et al. [101] reported that the functionalization of an SPI film with TiO_2 promoted a decrease in water vapor (from 5.43 to 4.62 g·mm·m^{-2}·day^{-1}·kPa^{-1}) and oxygen (from 0.470 to 0.110 g·cm^{-2}·day^{-1}) permeability, as well as an increase in tensile strength (from 6.6683 to 14.5642 MPa) in a TiO_2 concentration-dependent response. They argue that the presence of TiO_2 in protein structure significantly changes the hydrophilic nature of the film, due to the stable covalent (Si–O–C, Ti–O–C, and Si–O–Ti) and non-covalent (hydrogen bonds and Van der Waals forces) interactions between TiO_2 and SPI. Moreover, the hybrid film exhibited antimicrobial effects against *E. coli* (inhibition zone by agar test diffusion assay of 27.34 mm). Wang et al. [23] demonstrated the bactericidal efficiency of an SPI–TiO_2 hybrid film under UV-light (at 365 nm during two hours) against *E. coli* (reduction of 71.01% of viable cells) and *S. aureus* (reduction of 88.94% of viable cells), which was associated with the synergistic antimicrobial effect between TiO_2 and β-conglycinin and glycinin peptides present in the SPI [107].

Additionally, Wang et al. [19] informed that TiO_2 incorporation in an SPI film positively influences its tensile strength (90.79% higher than control). On the other hand, the addition of nano-TiO_2 reduced the flexibility (70.21% less than control), and water vapor (65.67% less than control), and oxygen (46.50% less than control) permeability in comparison with control groups. This was due to the strong hydrogen bonds formed between the two main components, which could prevent water and oxygen from diffusing through the films. The reduction in flexibility values could be associated with a collapse of the crystalline structure of the hybrid material by the formation of aggregates by an excess of TiO_2.

The reported application of SPI–TiO_2 hybrid film includes fruit preservation and water-dye degradation. Zhang et al. [105] reported that SPI–TiO_2 hybrid film was effective to extend the shelf life of strawberries stored at 4 °C up to 8 days without significant weight losses (<17.3%) and color changes with stable microbial quality in comparison with the uncoated fruits. Similar trends were reported in grapes coated with an SPI–TiO_2 hybrid film by Hoseiniyan et al. [95], who reported that coated grapes exhibited good performance during cold storage (31 days at 4 °C) without significant effects in the total soluble solids, titratable acidity, and weight losses. The hybrid film prevents the fungal infection of the fruits, and the coated fruits also had a good appearance and marketability compared with the uncoated fruits.

In summary, the incorporation of TiO_2 into SPI significantly improved its physicochemical properties and exhibited good fruit preservation performance.

5.4.2. Other Applications of Soy Protein Isolate–TiO_2 Hybrid Composite

Calza et al. [108] fabricated a system composed of soybean peroxidase and TiO_2 nanoparticles for environmental remediation purposes (Table 7). They informed that the hybrid material effectively remove orange II dye (100%) and carbamazepine (100%) drug from aqueous solutions after 60 min of exposure compared with the soybean peroxidase structure (<80% and <10%, respectively, after 120 min of exposure), which was associated with the synergistic properties of peroxidase and TiO_2. Further studies are needed to understand the removal and degradation mechanism of soybean peroxidase–TiO_2, which could be used as an alternative for wastewater treatment.

5.5. *Other Proteins Functionalized with TiO_2*

Table 8 lists various non-conventional proteins functionalized with TiO_2, such as zein, keratin, sodium caseinate, lactoferrin, and sesame, to enhance their physicochemical properties, where the most common method for their preparation is evaporative casting for films and freeze-drying for hydrogels and scaffolds. Furthermore, the nanoparticles used are commercially available with sizes ranging from 10 to 200 nm in its anatase phase, and in some cases in its rutile phase, using concentrations ranging from 0.5% to 10% in weight of total solid content.

Table 8. Effect of TiO_2 incorporation on non-conventional protein-based materials.

Application	Method/Presentation	* Composition	TiO_2 Specifications	Relevant Results	Ref.
Food and non-food packaging	Evaporative casting/Film	Zein (13.5% *w/v*), glycerol:PEG 600 (3.3% *w/w*)	Commercial SM: Hydrothermal (TiO_2:SiO_2): 1.5% *w/v* Size: 100–180 nm	TiO_2 improved the mechanical, thermal, and water-related properties of zein film.	[24]
Food packaging	Evaporative casting/Nanofibers	Zein (3 g 10 mL^{-1} of 70% aqueous ethanol)	Commercial (TiO_2:SiO_2): 5% *w/w* Size: <25 nm CP: Anatase Purity: 99.7%	Coated fruits extend their shelf life.	[33]
Food and non-food packaging	Evaporative casting/Film	Zein: sodium alginate (90:10), betanin (1%)	Commercial (TiO_2): 0.5% *w/w* Size: 10–25 nm	The hybrid film exhibited antimicrobial activity.	[109]
Food and non-food packaging	Evaporative casting/Film	Sodium caseinate (8 g 100 mL^{-1}), guar gum (0.3% *w/w*), CEO (2% *w/w*)	Commercial (TiO_2): 1% *w/w* Size: 10–25 nm CP: Anatase Purity: >99%	The hybrid film exhibited antimicrobial activity.	[11]
Food and non-food packaging	Evaporative casting/Film	Sodium caseinate (2.5% *w/w*), glycerol (2% *w/w*)	Commercial (P25) (TiO_2): 0.5% *w/w*	TiO_2 improved the mechanical, thermal, and water-related properties of the film.	[30]

Table 8. *Cont.*

Application	Method/Presentation	* Composition	TiO_2 Specifications	Relevant Results	Ref.
Food and non-food Packaging	Evaporative casting/Films	Feather keratin (1.2 g), PVA (13.33 g)	Commercial (P25) (TiO_2): 3% *w/w* Size: 60 nm CP: Anatase Purity: 99.8%	The hybrid material exhibited improved physicochemical properties.	[110]
Food and non-food Packaging	Catalyst curing/Composite	Raw wool keratin (350 g/m^2), BTCA (12.6%)	Commercial (P25) (TiO_2): 0.6 g·L^{-1} Size: 21 nm CP: Anatase-Rutile Crystalline structure	The hybrid material showed an improved UV-protective effect.	[22]
Environmental remediation	Evaporative casting/Film	Sesame protein (3 g 100 mL^{-1}), glycerol (30% in total solid content)	Commercial (P25) (TiO_2): 3% *w/w* Size: 21 nm CP: Anatase-Rutile Crystalline structure	The hybrid film exhibited photocatalytic activity against methylene blue.	[16]
Environmental remediation	Hydrogel synthesis/Hydrogel	Keratin (1% *w/v*)	Commercial (P25) (TiO_2): 10 *w/w* CP: Anatase-Rutile	Hybrid hydrogel effectively removes trimethoprim from wastewater.	[111]
Environmental remediation	Electrospinning/ Nanofibers	Keratin–PLA–TiO_2 mass ratio of 33:33:33	Commercial (P25) CP: Anatase	The hybrid nanofibers effectively remove methylene blue dye from the aqueous solution.	[112]
Environmental remediation	Biometic/Microspheres	NI	Anatase	The hybrid composite showed good photocatalytic properties again or dye yellow and blue acid dyes.	[113]
Biomedical	Freeze dried/Scaffolds	Silk fibroin (2% *w/v*), F (2% *v/v*)	Commercial (P25) (TiO_2): 15 *w/w*	SF–TiO_2:F exhibited biocompatibility and improved mechanical properties.	[114]
Biomedical	Freeze-dried/Scaffolds	Silk fibroin (2.5% *w/v*), chitin (2.5% *w/v*), glutaraldehyde (0.25% *v/v*)	Commercial (P25) (TiO_2): 1.5% *w/w* Size: 10–15 nm CP: Anatase Purity: >99%	Hybrid material exhibited antimicrobial activity, also it is biocompatible and biodegradable.	[115]
Biomedical	Dip-coating/Coating	Lactoferrin (0.2 mg·mL^{-1}), collagen (0.2 mg·mL^{-1})	SM: Sol-gel (TiO_2): NI Size: 200 nm CP: Anatase Crystalline structure	The hybrid coating showed enhanced biocompatibility with MG-6e cells.	[116]

* Material composition was based on the best-reported results. NI: No information; CEO: cumin essential oil; PVA: polyvinyl alcohol; PEG: polyethylene glycol; BTCA: 1,2,3,4-butane tetracarboxylic acid; PLA: poly(Lactic acid); SM: synthesis method; (TiO_2): concentration of titanium dioxide; CP: crystallite phase.

5.5.1. Packaging Applications of Non-Conventional Proteins Functionalized with TiO_2

Table 8 lists reports on the use of non-conventional protein materials functionalized with TiO_2 for food and non-food packaging development. Kadam et al. [24] evaluated the effect of TiO_2:SiO_2 nanoparticles incorporation on the thermal and mechanical properties of a cast zein film. They reported that mechanical properties (tensile strength) of the hybrid film were enhanced; however, its flexibility was reduced two-fold compared with zein film, possibly associated with the formation of TiO_2 aggregates. Furthermore, the water contact angle, water vapor permeability, and thermal properties of the hybrid film were improved by the addition of inorganic nanoparticles, associated with the interaction between zein and TiO_2:SiO_2, which promotes a stable and strong hydrogen bonds formation. Similarly, Amjadi et al. [109] made zein–sodium alginate (90:10) film functionalized with TiO_2–betanin (0.5%:1%) nanoparticles and informed that the hybrid film exhibited antioxidant properties (by the presence of bioactive compounds in betanin) and high antimicrobial effects (by agar test diffusion assay) against *E. coli* (15.4 mm of inhibition zone) and *S. aureus* (16.9 mm of inhibition zone), which was attributed to the antimicrobial properties of TiO_2. Moreover, Böhmer-Maas et al. [33] developed a zein–TiO_2 nanofiber as an ethylene absorber for cherry tomatoes preservation (25 °C). They reported that coated fruits with the hybrid film exhibited less ethylene concentration (9.38 $\mu g \cdot L^{-1} \cdot g^{-1} \cdot h^{-1}$) than those coated with a zein film (10.27 $\mu g \cdot L^{-1} \cdot g^{-1} \cdot h^{-1}$), which permits extended the shelf life of cherry tomatoes up to 22 days. According to the authors, the ethylene degradation occurs by the oxidation of ethylene into CO_2 and water by the OH radicals and reactive oxygen species generated by the photocatalytic ability of TiO_2.

Montes-de-Oca-Ávalos et al. [30] investigated the effect of TiO_2 incorporation on the physicochemical properties of a sodium caseinate film. They informed that mechanical, thermal, water vapor permeability characteristics of the caseinate film were improved in a TiO_2 concentration-dependent way, associated with good dispersion of TiO_2 through the film polymeric matrix. According to the authors, the presence of TiO_2 avoids protein agglomeration due to the stable hydrogen bond formation. Additionally, Alizadeh-Sani et al. [11] informed that a sodium caseinate–guar gum film functionalized with TiO_2 (1% *w/w*) and cumin essential oil (2% *w/w*) showed remarkable antimicrobial activity against *L. monocytogenes* (16 mm of inhibition zone), *S. aureus* (15 mm of inhibition zone), *E. coli* O157:H7 (14 mm of inhibition zone), *S. enteritidis* (12 mm of inhibition zone) in a strain-dependent manner. These results were associated with the cell wall differences between bacteria (outer membrane) and the synergistic antimicrobial effect among TiO_2 and cumin essential oil. Moreover, the water vapor permeability, tensile strength, and flexibility of the combined film were improved by a synergistic effect of TiO_2 and cumin essential oil.

Additionally, Montazer et al. [22] informed that the incorporation of TiO_2 in a wool keratin film stabilized by butane tetracarboxylic acid (BTCA) exhibited excellent UV-barrier properties related to the C–N and N–H bonds promoted for TiO_2 and BTCA interactions, with an optimum concentration of 0.6 $g \cdot L^{-1}$ and 12.94% *w/v*. Similarly, Wu et al. [110], who informed that thermal stability, mechanical resistance, and water vapor permeability of the keratin–tris film were improved by its functionalization with TiO_2 that may act as a physical cross-linker agent.

According to evidence, functionalization of non-conventional proteins like zein, keratin, and sodium caseinate with TiO_2 nanoparticles exhibited interesting properties for food and non-food packaging development.

5.5.2. Environmental Applications of Non-Conventional Proteins Functionalized with TiO_2

Usage of zein, keratin, and sesame proteins as a supporting material of TiO_2 for the removal and degradation of water pollutants have been explored (Table 8). Babitha and Korrapati [113] made mesoporous microspheres formed by zein and TiO_2 as an alternative for acid yellow (AY110) and acid blue (AB113) dyes decolorization under UV-light irradiation. They reported that the hybrid microspheres (1 $mg \cdot mL^{-1}$) showed a dye removal efficiency of 96% and 89% in AY110 and AB113, respectively, at lower dye concentration (10 $mg \cdot L^{-1}$) but decreased at higher concentrations (100 $mg \cdot L^{-1}$), which was associated with the saturation of active sites into the hybrid matrix.

Additionally, Villanueva et al. [111] fabricated a hydrogel combining keratin (from cow's horn) and TiO_2 to remove trimethoprim from wastewater. They reported that the hybrid material exhibited good degradation efficiency (>95%) against antibiotic removal from aqueous solution in a TiO_2 dose-dependent response, with an optimum TiO_2 concentration of 10% *w/w* with performance up to four consecutive cycles (90%). It was associated with the swelling and adsorptive abilities of the hybrid film and to the presence of active sites on the catalyst surface due to the strong attachment between keratin and TiO_2 through covalent and non-covalent interactions. Moreover, Siriorn and Jatuphorn [112] reported that a chicken feather keratin–poly(lactic acid)–TiO_2 nanofibers (0.05 g) effectively remove methylene blue (90%) dye from aqueous solution (5×10^{-6} M) under visible light due to the improved adsorptive properties of the hybrid nanofibers.

Fathi et al. [16] made a sesame protein isolate film functionalized with TiO_2 for water-dye removal purposes. They reported that the hybrid film (64 cm^2) effectively degraded 76% of methylene blue dye (10 $mg \cdot mL^{-1}$) under UV-light irradiation after 120 min of exposure. Moreover, the hybrid material exhibited enhanced water vapor permeability, water resistance, water contact angle, and mechanical strength in a TiO_2 dose-dependent response with an optimum TiO_2 concentration of 3% *w/w* associated with the interaction chemical and physical interactions between sesame protein and TiO_2. On the other hand, the morphological studies through scanning electron microscopy revealed that a high concentration of TiO_2 exhibited an inhomogeneous dispersion, causing aggregations in the protein matrix that negatively affects its functionality.

To summarize, non-conventional proteins like zein, keratin, and sesame functionalized with TiO_2 nanoparticles could be a viable, low-cost, and efficient alternative for environmental applications as photocatalysts for wastewater treatment.

5.5.3. Other Applications of Non-Conventional Proteins Functionalized with TiO_2

Other potential uses of non-conventional proteins functionalized with TiO_2 include bone regeneration, antimicrobial activity, and textiles (Table 8). Johari et al. [114] made a fluorated silk fibroin–TiO_2 hybrid scaffold for bone tissue engineering with non-toxic effects in human osteoblast cells (SaOS-2) and suitable cell attachment and spreading on the hybrid material, which was associated with the fluoridation of TiO_2 nanoparticles (TiO_2–F). Moreover, the hybrid scaffold exhibited good porosity (200 to 500 μm), mechanical resistance (tensile strength of 1.7 MPa), and adequate biodegradation rate (from 1% to 5% of weight loss in 30 days) in a TiO_2 dose-dependent response due to the formation of Ti–O–C bonds and the partial substitution of OH groups present in the TiO_2 surface by fluorine anions, that significantly increase the functional properties of TiO_2. On the other hand, with high amounts of TiO_2 (>15%), some agglomerates could appear that negatively affect the technological properties of the hybrid scaffold.

Mehrabani et al. [115] informed that a chitin–fibroin–TiO_2 hybrid composite did not show cytotoxic effects on a human Caucasian fetal foreskin fibroblast cell line at low TiO_2 concentrations (<1.5% *w/w*). Nonetheless, the hybrid material exhibited a porosity of 94%, a density of 3118 $mg \cdot mL^{-1}$, and water resistance with a swelling degree of 93% after 24 h. In addition, it showed antimicrobial properties against *E. coli*, *S. aureus*, and *C. albicans*, which are suitable for the development of wound-healing materials. According to Feng et al. [117], incorporation of TiO_2 into fibroin (mostly α-helix) matrix promotes structural changes that permit a strong interaction with the β-sheets changing from typical silk I to Silk II structure in a TiO_2-dependent manner, attributed to the presence of hydroxyl groups on the TiO_2. The enhanced properties of fibroin could be related to the conformational structure. On the other hand, the authors reported that a high concentration of TiO_2 might negatively affect the mechanical properties of the hybrid material associated with the damage of its microscopic structure mainly by the formation of TiO_2 agglomerates, and possibly to the extra water used for the preparation of the hybrid material.

Kazek-Kesik et al. [116] coated a lactoferrin–collagen composite on titanium alloys for bone replacement. It was found that the presence of lactoferrin and TiO_2 enhanced osteoblast-like effect on MG-63 cells after seven days of evaluation in comparison with collagen-treated cells, mainly by the ability of both components to promote cell adhesion.

According to the evidence, the functionalization of non-conventional proteins with TiO_2 nanoparticles exhibited interesting properties and applications. However, further studies are needed to validate their potential uses.

6. Disadvantages of Protein–TiO_2 Hybrid Composites and Perspectives

Despite the observation that protein–TiO_2 hybrid composites exhibited excellent technological and functional properties with great potential to be used in several applications, it is necessary to evaluate the safe use and implementation of this kind of hybrid composites, mainly due to the presence of TiO_2 in their composition.

In this context, it has been reported that pure TiO_2 exhibited toxicological and adverse effects in cell lines (HeLa and HaCaT), proteins (microtubule and bovine serum albumin), and animal models

(Sprague–Dawley rats, Wistar rats, and mussel *Mytilus coruscus*) in a concentration-dependent response, typically at doses ranging from 0.4 to 100 mg·mL^{-1} with direct application [118–123]. Nonetheless, the tested concentrations of TiO_2 in these works were higher than the recommended safe usage (<1% by weight) by international regulations in the use of TiO_2 as a food additive [124].

However, the amount of TiO_2 used as a functional agent to develop protein–TiO_2 hybrid composites ranges from 0.003 to 1 mg·mL^{-1}, depending on its application. For example: in food packaging materials manufacturing, the amount average of TiO_2 employed is 0.28 mg·mL^{-1}, while for packaging materials with non-food purposes it is 0.85 mg·mL^{-1}. Moreover, for the development of scaffolds, dental implants, and wound-healing materials, the average amount of TiO_2 is 0.23 and 0.9 mg·mL^{-1} for making hybrid materials for environmental remediation.

According to Xu et al. (2017) [123], the interaction between protein structure and TiO_2 plays a critical role in the safe use of these materials, which usually depends on the new properties of each hybrid composite and the used concentration of TiO_2 [125]. In this sense, there are a few reports on the toxicity status of protein–TiO_2 hybrid composites, which reported no toxicological or adverse effects on their use, associated with the low concentration of TiO_2 used for the functionalization of protein-based materials. However, most of the published reports cited in this document focused on in vitro evaluations. Therefore, further studies are needed to evaluate the possible human health and environmental risks on the usage of these hybrid composites.

7. Concluding Remarks

Significant evidence indicates that functionalization of protein-based materials by adding TiO_2 nanoparticles is a feasible approach to improve their thermal, mechanical, optical, water-resistance, gas-barrier, and adsorptive properties. The evaporative casting method is one of the most common procedures for the preparation of protein–TiO_2 hybrid films and coating and freeze-drying for hydrogels and scaffolds, using commercial TiO_2 with a particle size ranging from 10 to 200 nm (the most frequently used is 10–25 nm in size) in its anatase phase with a crystalline structure.

Protein–TiO_2 hybrid composites are an active research area for developing eco-friendly and active food and non-food packaging materials with antimicrobial and UV-protective effects. Furthermore, they are attractive and biocompatible materials to fabricate wound-healing patches, tissue engineering scaffolds, or biosensors for biomedical applications.

On the other hand, although the functionalization of protein-based materials with TiO_2 offers significant advantages, some limitations have been reported, especially those associated with the concentration of TiO_2. Higher concentrations of TiO_2 could promote an inhomogeneous dispersion through the polymeric matrix, forming agglomerates that negatively affected the technological and functional properties of the hybrid material, particularly in flexibility and transparency. Likewise, the preparation method could negatively influence the properties of the hybrid material, associated with the physical and chemical interactions between components. For example, if there was no proper mixing ratio between protein and TiO_2, a saturation of the available functional groups in the polymeric matrix can affect the physicochemical properties of the film. Additionally, other possible limitations of the protein–TiO_2 hybrid composites could be related to the type and source of protein and its possible structural changes by the presence of TiO_2 and its stability for diverse applications.

There are some challenges to be achieved for industrial applications; one of the most important is to obtain the correct amounts of protein and TiO_2 nanoparticles because different uses require different formulations with desirable properties. For example, the shelf life of climacteric fruits depends on the correct exchange of oxygen, carbon dioxide, and water vapor permeability. Meanwhile, products with high amounts of lipids require UV-protective effects to prevent their oxidation. On the other hand, wound-healing materials should exhibit high water and mechanical resistance but correct gas exchange, high adherence, and antimicrobial properties. Moreover, standardized protocols for their preparation are needed for industrial-scale implementation. It is also necessary to carry out in vivo tests to evaluate the possible human health and environmental risks on the usage and safe implementation of these hybrid composites in diverse applications. Therefore, further research efforts should be dedicated to solving these challenges.

Author Contributions: Conceptualization, L.M.A.-E., Z.V.-d.l.M., D.A.L.-d.l.M., A.P.-L., and E.M.G.; methodology, L.M.A.-E., Z.V.-d.l.M., N.R.-B., T.S.-C., K.N., D.A.L.-d.l.M, A.P.-L., and E.M.-G.; investigation, L.M.A.-E., Z.V.-d.l.M., N.R.-B., T.S.-C., K.N., D.A.L.-d.l.M., A.P.-L., and E.M.-G.; writing–original draft preparation, L.M.A.-E., Z.V.-d.l.M., N.R.-B., T.S.-C., K.N., D.A.L.-d.l.M., A.P.-L., and E.M.-G.; writing–review and editing, L.M.A.-E., Z.V.-d.l.M., D.A.L.-d.l.M., A.P.-L., and E.M.-G.; supervision, L.M.A.-E., Z.V.-d.l.M., D.A.L.-d.l.M., A.P.-L., and E.M.-G. All authors have read and agreed to the published version of the manuscript.

Funding: This research received no external funding.

Acknowledgments: The authors gratefully acknowledge the financial support from a scholarship (702634) from CONACYT-Mexico, as well as Acoyani Garrido-Sandoval for proofreading the English language of this research.

Conflicts of Interest: The authors declare no conflict of interest.

References

1. Riahi, Z.; Priyadarshi, R.; Rhim, J.; Bagheri, R. Gelatin-based functional films integrated with grapefruit seed extract and TiO_2 for active food packaging applications. *Food Hydrocoll.* **2021**, *112*, 106314. [CrossRef]
2. Avramescu, S.M.; Butean, C.; Popa, C.V.; Ortan, A.; Moraru, I.; Temocico, G. Edible and functionalized films/coatings-performances and perspectives. *Coatings* **2020**, *10*, 687. [CrossRef]
3. Lin, D.; Yang, Y.; Wang, J.; Yan, W.; Wu, Z.; Chen, H.; Zhang, Q.; Wu, D.; Qin, W.; Tu, Z. Preparation and characterization of TiO_2–Ag loaded fish gelatin-chitosan antibacterial composite film for food packaging. *Int. J. Biol. Macromol.* **2020**, *154*, 123–133. [CrossRef] [PubMed]
4. El-wakil, N.A.; Hassan, E.A.; Abou-zeid, R.E.; Dufresne, A. Development of wheat gluten/nanocellulose/ titanium dioxide nanocomposites for active food packaging. *Carbohydr. Polym.* **2015**, *124*, 337–346. [CrossRef] [PubMed]
5. Park, J.E.; Park, I.; Neupane, M.P.; Bae, T.; Lee, M. Effects of a carbon nanotube-collagen coating on a titanium surface on osteoblast growth. *Appl. Surf. Sci.* **2014**, *292*, 828–836. [CrossRef]
6. Al Zoubi, W.; Kamil, M.P.; Fatimah, S.; Nisa, N.; Ko, Y.G. Recent advances in hybrid organic-inorganic materials with spatial architecture for state-of-the-art applications. *Prog. Mater. Sci.* **2020**, *112*, 100663. [CrossRef]
7. Zhao, Y.; Liu, J.; Zhang, M.; He, J.; Zheng, B.; Liu, F.; Zhao, Z.; Liu, Y. Use of silver nanoparticle–gelatin/alginate scaffold. *Coatings* **2020**, *10*, 948. [CrossRef]
8. Anaya-Esparza, L.M.; Villagrán-de la Mora, Z.; Ruvalcaba-Gómez, J.M.; Romero-Toledo, R.; Sandoval-Contreras, T.; Aguilera-Aguirre, S.; Montalvo-González, E.; Pérez-Larios, A. Use of titanium dioxide (TiO_2) nanoparticles as reinforcement agent of polysaccharide-based materials. *Processes* **2020**, *8*, 1395. [CrossRef]
9. Anaya-Esparza, L.M.; Ruvalcaba-Gómez, J.M.; Maytorena-Verdugo, C.I.; González-Silva, N.; Romero-Toledo, R.; Aguilera-Aguirre, S.; Pérez-Larios, A.; Montalvo-González, E. Chitosan–TiO_2: A versatile hybrid composite. *Materials* **2020**, *13*, 811. [CrossRef]
10. Omar, N.; Selvami, S.; Kaisho, M.; Yamada, M.; Yasui, T.; Fukumoto, M. Deposition of titanium dioxide coating by the cold-spray process on annealed stainless steel substrate. *Coatings* **2020**, *10*, 991. [CrossRef]
11. Alizadeh-Sani, M.; Rhim, J.W.; Azizi-Lalabadi, M.; Hemmati-Dinarvand, M.; Ehsani, A. Preparation and characterization of functional sodium caseinate/guar gum/TiO_2/cumin essential oil composite film. *Int. J. Biol. Macromol.* **2020**, *145*, 835–844. [CrossRef] [PubMed]
12. de Fonseca, J.M.; Valencia, G.A.; Soares, L.S.; Dotto, M.E.R.; Campos, C.E.M.; de Moreira, R.F.P.M.; Fritz, A.R.M. Hydroxypropyl methylcellulose–TiO_2 and gelatin–TiO_2 nanocomposite films: Physicochemical and structural properties. *Int. J. Biol. Macromol.* **2020**, *151*, 944–956. [CrossRef] [PubMed]
13. Anaya-Esparza, L.M.; Montalvo-González, E.; González-Silva, N.; Méndez-Robles, M.D.; Romero-Toledo, R.; Yahia, E.M.; Pérez-Larios, A. Synthesis and characterization of TiO_2–ZnO–MgO mixed oxide and their antibacterial activity. *Materials* **2019**, *12*, 698. [CrossRef] [PubMed]
14. Anaya-Esparza, L.M.; González-Silva, N.; Yahia, E.M.; González-Vargas, O.A.; Montalvo-González, E.; Pérez-Larios, A. Effect of TiO_2–ZnO–MgO mixed oxide on microbial growth and toxicity against *Artemia salina*. *Nanomaterials* **2019**, *9*, 992. [CrossRef] [PubMed]

15. Ortelli, S.; Costa, A.L. Insulating thermal and water-resistant hybrid coating for fabrics. *Coatings* **2020**, *10*, 72. [CrossRef]
16. Fathi, N.; Almasi, H.; Pirouzifard, M.K. Sesame protein isolate based bionanocomposite films incorporated with TiO_2 nanoparticles: Study on morphological, physical and photocatalytic properties. *Polym. Test.* **2019**, *77*, 105919. [CrossRef]
17. Li, Y.; Jiang, Y.; Liu, F.; Ren, F.; Zhao, G.; Leng, X. Fabrication and characterization of TiO_2/whey protein isolate nanocomposite film. *Food Hydrocoll.* **2011**, *25*, 1098–1104. [CrossRef]
18. Farshchi, E.; Pirsa, S.; Roufegarinejad, L.; Alizadeh, M.; Rezazad, M. Photocatalytic/biodegradable film based on carboxymethyl cellulose, modified by gelatin and TiO_2–Ag nanoparticles. *Carbohydr. Polym.* **2019**, *216*, 189–196. [CrossRef]
19. Wang, Z.; Zhang, N.; Wang, H.Y.; Sui, S.Y.; Sun, X.X.; Ma, Z.S. The effects of ultrasonic/microwave assisted treatment on the properties of soy protein isolate/titanium dioxide films. *LWT-Food Sci. Technol.* **2014**, *57*, 548–555. [CrossRef]
20. Boughriba, S.; Souissi, N.; Jridi, M.; Li, S.; Nasri, M. Thermal, mechanical and microstructural characterization and antioxidant potential of *Rhinobatos cemiculus* gelatin films supplemented by titanium dioxide doped silver nanoparticles. *Food Hydrocoll.* **2020**, *103*, 105695. [CrossRef]
21. Vejdan, A.; Ojagh, S.M.; Adeli, A.; Abdollahi, M. Effect of TiO_2 nanoparticles on the physico-mechanical and ultraviolet light barrier properties of fish gelatin/agar bilayer film. *LWT Food Sci. Technol.* **2016**, *71*, 88–95. [CrossRef]
22. Montazer, M.; Pakdel, E.; Moghadam, M.B. Nano titanium dioxide on wool keratin as UV absorber stabilized by butane tetra carboxylic acid (BTCA): A statistical prospect. *Fibers Polym.* **2010**, *11*, 967–975. [CrossRef]
23. Wang, S.Y.; Zhu, B.B.; Li, D.Z.; Fu, X.Z.; Shi, L. Preparation and characterization of TiO_2/SPI composite film. *Mater. Lett.* **2012**, *83*, 42–45. [CrossRef]
24. Kadam, D.M.; Thunga, M.; Srinivasan, G.; Wang, S.; Kessler, M.R.; Grewell, D.; Yu, C.; Lamsal, B. Effect of TiO_2 nanoparticles on thermo-mechanical properties of cast zein protein films. *Food Packag. Shelf Life* **2017**, *13*, 35–43. [CrossRef]
25. He, Q.; Huang, Y.; Lin, B.; Wang, S. A nanocomposite film fabricated with simultaneously extracted protein-polysaccharide from a marine alga and TiO_2 nanoparticles. *J. Appl. Phycol.* **2017**, *29*, 1541–1552. [CrossRef]
26. He, Q.; Zhang, Y.; Cai, X.; Wang, S. Fabrication of gelatin–TiO_2 nanocomposite film and its structural, antibacterial and physical properties. *Int. J. Biol. Macromol.* **2016**, *84*, 153–160. [CrossRef]
27. Fan, X.; Chen, K.; He, X.; Li, N.; Huang, J.; Tang, K.; Li, Y.; Wang, F. Nano-TiO_2/collagen-chitosan porous scaffold for wound repairing. *Int. J. Biol. Macromol.* **2016**, *91*, 15–22. [CrossRef]
28. Alizadeh Sani, M.; Ehsani, A.; Hashemi, M. Whey protein isolate/cellulose nanofibre/TiO_2 nanoparticle/rosemary essential oil nanocomposite film: Its effect on microbial and sensory quality of lamb meat and growth of common foodborne pathogenic bacteria during refrigeration. *Int. J. Food Microbiol.* **2017**, *251*, 8–14. [CrossRef]
29. Feng, Z.; Li, L.; Wang, Q.; Wu, G.; Liu, C.; Jiang, B.; Xu, J. Effect of antioxidant and antimicrobial coating based on whey protein nanofibrils with TiO_2 nanotubes on the quality and shelf life of chilled meat. *Int. J. Mol. Sci.* **2019**, *20*, 1184. [CrossRef]
30. Montes-de-Oca-Ávalos, J.M.; Altamura, D.; Jorge, R. Relationship between nano/microstructure and physical properties of TiO_2–sodium caseinate composite films. *Food Res. Int.* **2018**, *105*, 129–139. [CrossRef]
31. Zolfi, M.; Khodaiyan, F.; Mousavi, M.; Hashemi, M. The improvement of characteristics of biodegradable films made from kefiran-whey protein by nanoparticle incorporation. *Carbohydr. Polym.* **2014**, *109*, 118–125. [CrossRef] [PubMed]
32. Zink, J.; Wyrobnik, T.; Prinz, T.; Schmid, M. Physical, chemical and biochemical modifications of protein-based films and coatings: An extensive review. *Int. J. Mol. Sci.* **2016**, *17*, 1376. [CrossRef] [PubMed]
33. Böhmer-Maas, B.W.; Martins, L.; Murowaniecki, D.; Zavareze, R.; Carlos, R. Photocatalytic zein–TiO_2 nanofibers as ethylene absorbers for storage of cherry tomatoes. *Food Packag. Shelf Life* **2020**, *24*, 100508. [CrossRef]
34. Li, N.; Fan, X.; Tang, K.; Zheng, X.; Liu, J.; Wang, B. Nanocomposite scaffold with enhanced stability by hydrogen bonds between collagen, polyvinyl pyrrolidone and titanium dioxide. *Coll. Surf. B* **2016**, *140*, 287–296. [CrossRef] [PubMed]

35. Pop, O.L.; Pop, C.R.; Dufrechou, M.; Vodnar, D.C.; Socaci, S.A.; Dulf, F.V.; Minervini, F.; Suharoschi, R. Edible films and coatings functionalization by probiotic incorporation: A review. *Polymers* **2020**, *12*, 12. [CrossRef]
36. Acquah, C.; Zhang, Y.; Dubé, M.A.; Udenigwe, C.C. Formation and characterization of protein-based films from yellow pea (*Pisum sativum*) protein isolate and concentrate for edible applications. *Curr. Res. Food Sci.* **2020**, *2*, 61–69. [CrossRef]
37. Agudelo-Cuartas, C.; Granda-restrepo, D.; Sobral, P.J.A.; Hernandez, H. Characterization of whey protein-based fi lms incorporated with natamycin and nanoemulsion of α-tocopherol. *Heliyon* **2020**, *6*, e03809. [CrossRef]
38. Guo, X.; Ren, C.; Zhang, Y.; Cui, H.; Shi, C. Stability of zein-based films and their mechanism of change during storage at different temperatures and relative humidity. *J. Food Process. Preserv.* **2020**, *44*, 1–10. [CrossRef]
39. Su, J.; Huang, Z.; Liu, K.; Fu, L.; Liu, H. Mechanical properties, biodegradation and water vapor permeability of blend films of soy protein isolate and poly(vinyl alcohol) compatibilized by glycerol. *Polym. Bull.* **2007**, *921*, 913–921. [CrossRef]
40. Wang, N.; Saleh, A.S.M.; Xiao, Z. Effect of protein aggregates on properties and structure of rice bran protein-based film at different pH. *J. Food Sci. Technol.* **2019**, *56*, 5116–5127. [CrossRef]
41. Mangavel, C.; Rossignol, N.; Perronnet, A.; Barbot, A.; Gueguen, J. Properties and microstructure of thermo-pressed Wheat gluten films: A comparison with cast films. *Biomacromolecules* **2004**, *5*, 1596–1601. [CrossRef] [PubMed]
42. Alexandre, E.M.C.; Lourenço, R.V.; Bittante, A.M.Q.B.; Moraes, I.C.F.; do Amaral Sobral, P.J. Gelatin-based films reinforced with montmorillonite and activated with nanoemulsion of ginger essential oil for food packaging applications. *Food Packag. Shelf Life* **2016**, *10*, 87–96. [CrossRef]
43. Dias, J.R.; Baptista-silva, S.; De Oliveira, C.M.T.; Sousa, A.; Oliveira, A.L. In situ crosslinked electrospun gelatin nanofibers for skin regeneration. *Eur. Polym. J.* **2017**, *95*, 161–173. [CrossRef]
44. Rouse, J.G.; Van Dyke, M.E. A review of keratin-based biomaterials for biomedical applications. *Materials* **2010**, *3*, 999–1014. [CrossRef]
45. Hauzoukim; Swain, S.; Mohanty, B.; Hauzoukim, S.S.; Mohanty, B. Functionality of protein-based edible coating—Review. *J. Entomol. Zool. Stud.* **2020**, *8*, 1432–1440.
46. Kanmani, P.; Rhim, J.W. Physicochemical properties of gelatin/silver nanoparticle antimicrobial composite films. *Food Chem.* **2014**, *148*, 162–169. [CrossRef]
47. Bang, Y.J.; Shankar, S.; Rhim, J.W. In situ synthesis of multi-functional gelatin/resorcinol/silver nanoparticles composite films. *Food Packag. Shelf Life* **2019**, *22*, 100399. [CrossRef]
48. Cano, A.; Andres, M.; Chiralt, A.; González-Martinez, C. Use of tannins to enhance the functional properties of protein based films. *Food Hydrocoll.* **2020**, *100*. [CrossRef]
49. Alizadeh-Sani, M.; Moghaddas Kia, E.; Ghasempour, Z.; Ehsani, A. Preparation of active nanocomposite film consisting of sodium caseinate, ZnO nanoparticles and rosemary essential oil for food packaging applications. *J. Polym. Environ.* **2020**. [CrossRef]
50. Azevedo, V.M.; Dias, M.V.; de Siqueira Elias, H.H.; Fukushima, K.L.; Silva, E.K.; de Deus Souza Carneiro, J.; de Fátima Ferreira Soares, N.; Borges, S.V. Effect of whey protein isolate films incorporated with montmorillonite and citric acid on the preservation of fresh-cut apples. *Food Res. Int.* **2018**, *107*, 306–313. [CrossRef]
51. Pintado, C.M.B.S.; Ferreira, M.A.S.S.; Sousa, I. Properties of whey protein-based films containing organic acids and nisin to control *Listeria Monocytogenes*. *J. Food Prot.* **2009**, *72*, 1891–1896. [CrossRef] [PubMed]
52. Pluta-Kubica, A.; Jamróz, E.; Kawecka, A.; Juszczak, L.; Krzyściak, P. Active edible furcellaran/whey protein films with yerba mate and white tea extracts: Preparation, characterization and its application to fresh soft rennet-curd cheese. *Int. J. Biol. Macromol.* **2020**, *155*, 1307–1316. [CrossRef] [PubMed]
53. Arfat, Y.A.; Benjakul, S.; Prodpran, T.; Sumpavapol, P.; Songtipya, P. Physico-mechanical characterization and antimicrobial properties of fish protein isolate/fish skin gelatin-zinc oxide (ZnO) nanocomposite films. *Food Bioprocess Technol.* **2016**, *9*, 101–112. [CrossRef]
54. Salama, A. Soy protein acid hydrolysate/silica hybrid material as novel adsorbent for methylene blue. *Compos. Commun.* **2019**, *12*, 101–105. [CrossRef]

55. Asadzadeh, F.; Pirsa, S. Specific removal of nitrite from Lake Urmia sediments by biohydrogel based on isolated soy protein/tragacanth/mesoporous silica nanoparticles/lycopene. *Glob. Chall.* **2020**, *2000061*, 1–12. [CrossRef]
56. Hou, C.; Xu, Z.; Qiu, W.; Wu, R.; Wang, Y.; Xu, Q.; Liu, X.Y. A Biodegradable and stretchable protein-based sensor as artificial electronic skin for human motion detection. *Small* **2019**, *1805084*, 1–8. [CrossRef]
57. You, R.; Zhang, J.; Gu, S.; Zhou, Y.; Li, X.; Ye, D.; Xu, W. Regenerated egg white/silk fibroin composite films for biomedical applications. *Mater. Sci. Eng. C* **2017**, *79*, 430–435. [CrossRef]
58. Topoglidis, E.; Cass, A.E.G.; Gilardi, G.; Sadeghi, S.; Beaumont, N.; Durrant, J.R. Protein adsorption on nanocrystalline TiO_2 films: An immobilization strategy for bioanalytical devices. *Anal. Chem.* **1998**, *70*, 5111–5113. [CrossRef]
59. Ranjan, S.; Dasgupta, N.; Sudandiradoss, C.; Ramalingam, C.; Kumar, A. Titanium dioxide nanoparticle-protein interaction explained by docking approach. *Int. J. Nanomed.* **2018**, *13*, 47–50. [CrossRef]
60. Hashim, J.; Looney, L.; Hashmi, M.S.J. Metal matrix composites: Production by the stir casting method. *J. Mater. Process. Technol.* **1999**, *92*, 1–7. [CrossRef]
61. Asri, R.I.M.; Harun, W.S.W.; Hassan, M.A.; Ghani, S.A.C.; Buyong, Z. A review of hydroxyapatite-based coating techniques: Sol-gel and electrochemical depositions on biocompatible metals. *J. Mech. Behav. Biomed. Mater.* **2016**, *57*, 95–108. [CrossRef] [PubMed]
62. Jilani, A.; Abdel-wahab, M.S.; Hammad, A.H. Advance deposition techniques for thin film and coating. In *Modern Technologies for Creating the Thin-Film Systems and Coatings*; BoD–Books on Demand: Norderstedt, Germany, 2017; pp. 137–149.
63. Tang, X.; Yan, X. Dip-coating for fibrous materials: Mechanism, methods and applications. *J. Sol-Gel Sci. Technol.* **2017**, *81*, 378–404. [CrossRef]
64. Andres, C.M.; Kotov, N.A. Inkjet deposition of layer-by-layer assembled films. *J. Am. Chem. Soc.* **2010**, *132*, 14496–14502. [CrossRef] [PubMed]
65. Richardson, J.J.; Björnmalm, M.; Caruso, F. Technology-driven layer-by-layer assembly of nanofilms. *Science* **2015**, *348*, 411–422. [CrossRef]
66. Baudron, V.; Gurikov, P.; Smirnova, I.; Whitehouse, S. Porous starch materials via supercritical-and freeze-drying. *Gels* **2019**, *5*, 12. [CrossRef]
67. Liew, K.B.; Odeniyi, M.A.; Peh, K.K. Application of freeze-drying technology in manufacturing orally disintegrating films. *Pharm. Dev. Technol.* **2016**, *21*, 346–353. [CrossRef]
68. Matysiak, W.; Tanski, T.; Smok, W. Electrospinning as a versatile method of composite thin films fabrication for selected applications. *Solid State Phenom.* **2019**, *293*, 35–49. [CrossRef]
69. Amrollahi, P.; Krasinki, J.S.; Vaidyanathan, R.; Tayebi, L.; Vashaee, D. Electrophoretic deposition (EPD): Fundamentals and applications from nano- to micro-scale structures. Handb. *Nanoelectrochemistry* **2015**. [CrossRef]
70. Emregul, E.; Kocabay, O.; Derkus, B.; Yumak, T.; Emregul, K.C.; Sinag, A.; Polat, K. A novel carboxymethylcellulose-gelatin-titanium dioxide-superoxide dismutase biosensor; electrochemical properties of carboxymethylcellulose-gelatin-titanium dioxide-superoxide dismutase. *Bioelectrochemistry* **2013**, *90*, 8–17. [CrossRef]
71. Ferreira, R.; Padilla, R.; Urkasemsin, G.; Yoon, K.; Goeckner, K.; Hu, W.S.; Ko, C.C. Titanium-enriched hydroxyapatite-gelatin scaffolds with osteogenically differentiated progenitor cell aggregates for calvaria bone regeneration. *Tissue Eng. Part A* **2013**, *19*, 1803–1816. [CrossRef]
72. Gautam, R.K.; Kakatkar, A.S.; Karani, M.N. Development of protein-based biodegradable films from fish processing waste. *Int. J. Curr. Microbiol. Appl. Sci.* **2016**, *5*, 878–888. [CrossRef]
73. Nikpasand, A.; Parvizi, M.R. Evaluation of the Effect of titatnium dioxide nanoparticles/gelatin composite on infected skin wound healing; An animal model study. *Bull. Emerg. Trauma* **2019**, *7*, 366–372. [CrossRef] [PubMed]
74. Azizi-Lalabadi, M.; Alizadeh-Sani, M.; Divband, B.; Ehsani, A.; Julian, D. Nanocomposite films consisting of functional nanoparticles (TiO_2 and ZnO) embedded in 4A-Zeolite and mixed polymer matrices (gelatin and polyvinyl alcohol). *Food Res. Int.* **2020**, *137*, 109716. [CrossRef] [PubMed]
75. Azizi-Lalabadi, M.; Ehsani, A.; Ghanbarzadeh, B.; Divband, B. Polyvinyl alcohol/gelatin nanocomposite containing ZnO, TiO_2 or ZnO/TiO_2 nanoparticles doped on 4A-zeolite: Microbial and sensory qualities of packaged white shrimp during refrigeration. *Int. J. Food Microbiol.* **2020**, *312*, 108375. [CrossRef]

76. Nassiri, R.; Mohammady Nafchi, A. Antimicrobial and barrier properties of bovine gelatin films reinforced by nano TiO_2. *J. Chem. Health Risks* **2013**, *3*, 21–28.
77. Pirsa, S.; Farshchi, E.; Roufegarinejad, L. Antioxidant/antimicrobial film based on carboxymethyl cellulose/gelatin/TiO_2–Ag nanocomposite. *J. Polym. Environ.* **2020**. [CrossRef]
78. Vejdan, A.; Ojagh, S.M.; Abdollahi, M. Effect of gelatin/agar bilayer film incorporated with TiO_2 nanoparticles as a UV absorbent on fish oil photooxidation. *Int. J. Food Sci. Technol.* **2017**, *52*, 1862–1868. [CrossRef]
79. Urruela-Barrios, R.; Ramírez-Cedillo, E.; de León, A.D.; Alvarez, A.J.; Ortega-Lara, W. Alginate/gelatin hydrogels reinforced with TiO_2 and β-TCP fabricated by microextrusion-based printing for tissue regeneration. *Polymers* **2019**, *11*, 457. [CrossRef]
80. Lai, M.; Jin, Z.; Qiao, W. Surface immobilization of gelatin onto TiO_2 nanotubes to modulate osteoblast behavior. *Coll. Surf. B* **2017**, *159*, 743–749. [CrossRef]
81. Yan, Y.; Zhang, X.; Mao, H.; Huang, Y.; Ding, Q.; Pang, X. Hydroxyapatite/gelatin functionalized graphene oxide composite coatings deposited on TiO_2 nanotube by electrochemical deposition for biomedical applications. *Appl. Surf. Sci.* **2015**, *329*, 76–82. [CrossRef]
82. Hosokawa, A.; Kato, Y.; Terada, K. Imprinting on empty hard gelatin capsule shells containing titanium dioxide by application of the UV laser printing technique. *Drug Dev. Ind. Pharm.* **2014**, *9045*, 1047–1053. [CrossRef] [PubMed]
83. Hayajneh, M.T.; Almomani, M.; Al-daraghmeh, M. Enhancement the corrosion resistance of AISI-304 stainless steel by nanocomposite gelatin-titanium dioxide coatings. *Manuf. Technol.* **2019**, *19*, 759–766. [CrossRef]
84. Liu, B.; Xiao, J.; Xu, L.; Yao, Y.; Costa, B.F.O.; Domingos, V.F.; Ribeiro, E.S.; Shi, F.N.; Zhou, K.; Su, J.; et al. Gelatin-assisted sol-gel derived TiO_2 microspheres for hydrogen storage. *Int. J. Hydrog. Energy* **2015**, *40*, 4945–4950. [CrossRef]
85. Jao, D.; Xue, Y.; Medina, J.; Hu, X. Protein-based drug-delivery materials. *Materials* **2017**, *10*, 517. [CrossRef] [PubMed]
86. Alizadeh-Sani, M.; Khezerlou, A.; Ehsani, A. Fabrication and characterization of the bionanocomposite film based on whey protein biopolymer loaded with TiO_2 nanoparticles, cellulose nanofibers and rosemary essential oil. *Ind. Crops Prod.* **2018**, *124*, 300–315. [CrossRef]
87. Zhou, J.J.; Wang, S.Y.; Gunasekaran, S. Preparation and characterization of whey protein film incorporated with TiO_2 nanoparticles. *J. Food Sci.* **2009**, *74*. [CrossRef]
88. Alizadeh-Sani, M.; Mohammadian, E.; Julian, D. Eco-friendly active packaging consisting of nanostructured biopolymer matrix reinforced with TiO_2 and essential oil: Application for preservation of refrigerated meat. *Food Chem.* **2020**, *322*, 126782. [CrossRef]
89. Zhang, W.; Chen, J.; Chen, Y.; Xia, W.; Xiong, Y.L.; Wang, H. Enhanced physicochemical properties of chitosan/whey protein isolate composite film by sodium laurate-modified TiO_2 nanoparticles. *Carbohydr. Polym.* **2016**, *138*, 59–65. [CrossRef]
90. Zolfi, M.; Khodaiyan, F.; Mousavi, M.; Hashemi, M. Development and characterization of the kefiran-whey protein isolate-TiO_2 nanocomposite films. *Int. J. Biol. Macromol.* **2014**, *65*, 340–345. [CrossRef] [PubMed]
91. Gohargani, M.; Lashkari, H.; Shirazinejad, A. Study on biodegradable chitosan-whey protein-based film containing bionanocomposite TiO_2 and *Zataria multiflora* essential oil. *J. Food Qual.* **2020**, 8844167. [CrossRef]
92. Ortelli, S.; Malucelli, G.; Cuttica, F.; Blosi, M.; Zanoni, I.; Luisa, A. Coatings made of proteins adsorbed on TiO_2 nanoparticles: A new flame retardant approach for cotton fabrics. *Cellulose* **2018**, *25*, 2739–2749. [CrossRef]
93. Córdoba, L.C.; Hélary, C.; Montemor, F.; Coradin, T. Bi-layered silane–TiO_2/collagen coating to control biodegradation and biointegration of Mg alloys. *Mater. Sci. Eng. C* **2019**, *94*, 126–138. [CrossRef] [PubMed]
94. Truc, N.T.; Minh, H.H.; Khanh, L.L.; Thuy, V.M.; Van Toi, V.; Van Man, T.; Nam, H.C.N.; Quyen, T.N.; Hiep, N.T. Modification of type I collagen on TiO_2 surface using electrochemical deposition. *Surf. Coat. Technol.* **2018**, *344*, 664–672. [CrossRef]
95. Nojiri, T.; Chen, C.Y.; Kim, D.M.; Da Silva, J.; Lee, C.; Maeno, M.; Mcclelland, A.A.; Tse, B.; Nagai, S.I.; Hatakeyama, W.; et al. Establishment of perpendicular protrusion of type I collagen on TiO_2 nanotube surface as a priming site of peri-implant connective fibers. *J. Nanobiotechnol.* **2019**, *17*, 34. [CrossRef]
96. Bishal, A.K.; Sukotjo, C.; Jokisaari, J.R.; Klie, R.F.; Takoudis, C.G. Enhanced bioactivity of collagen fiber functionalized with room temperature atomic layer deposited titania. *ACS Appl. Mater. Interfaces* **2018**, *10*, 34443–34454. [CrossRef]

97. Vedhanayagam, M.; Anandasadagopan, S.; Nair, B.U.; Sreeram, K.J. Polymethyl methacrylate (PMMA) grafted collagen scaffold reinforced by PdO–TiO_2 nanocomposites. *Mater. Sci. Eng. C* **2019**, 110378. [CrossRef]
98. Erciyes, A.; Ocak, B. Physico-mechanical, thermal, and ultraviolet light barrier properties of collagen hydrolysate films from leather solid wastes incorporated with nano TiO_2. *Polym. Compos.* **2019**, *40*, 4716–4725. [CrossRef]
99. Luo, T.; Wan, X.J.; Jiang, S.X.; Zhang, L.Y.; Hong, Z.Q.; Liu, J. Preparation and photocatalytic performance of fibrous Tb^{3+}-doped TiO_2 using collagen fiber as template. *Appl. Phys. A Mater. Sci. Process.* **2018**, *124*, 304. [CrossRef]
100. Chen, H.; Liu, H.; Guo, Y.; Wang, B.; Wei, Y.; Zhang, Y.; Wu, H. Hierarchically ordered mesoporous TiO_2 nanofiber bundles derived from natural collagen fibers for lithium and sodium storage. *J. Alloys Compd.* **2017**, *731*, 844–852. [CrossRef]
101. Liu, Y.; Xu, L.; Li, R.; Zhang, H.; Cao, W.; Li, T.; Zhang, Y. Preparation and characterization of soy protein isolate films incorporating modified nano-TiO_2. *Int. J. Food Eng.* **2019**, *15*, 1–13. [CrossRef]
102. Malathi, A.N.; Kumar, N.; Nidoni, U.; Hiregoudar, S. Development of soy protein isolate films reinforced with titanium dioxide nanoparticles. *Int. J. Agric. Environ. Biotechnol.* **2017**, *10*, 141. [CrossRef]
103. Burgos, N.; Valdés, A.; Jiménez, A. Valorization of agricultural wastes for the production of protein-based biopolymers. *J. Renew. Mater.* **2016**, *4*, 165–177. [CrossRef]
104. Calva-Estrada, S.J.; Jiménez-Fernández, M.; Lugo-Cervantes, E. Protein-based films: Advances in the development of biomaterials applicable to food packaging. *Food Eng. Rev.* **2019**, *11*, 78–92. [CrossRef]
105. Zhang, Y.; Lin, S.; Lin, B.; Wang, W.; Wang, S. Studies on fresh-keeping strawberry using TiCVSPI composite film. *J. Chin. Inst. Food Sci. Technol.* **2015**, *15*, 120–125. [CrossRef]
106. Hoseiniyan, F.; Amiri, S.; Rezazadeh Bari, M.; Rezazad Bari, L.; Dodangeh, S. Effect of soy protein isolate and TiO_2 edible coating on quality and shelf-life of grapes varieties Hosseini and Ghezel Ozom. *FSCT* **2020**, *17*, 29–41.
107. Vasconcellos, F.C.S.; Woiciechowski, A.L.; Soccol, V.T.; Mantovani, D.; Soccol, C.R. Antimicrobial and antioxidant properties of β-conglycinin and glycinin from soy protein isolate. *Int. J. Curr. Microbiol. Appl. Sci.* **2014**, *3*, 144–157.
108. Calza, P.; Zacchigna, D.; Laurenti, E. Degradation of orange dyes and carbamazepine by soybean peroxidase immobilized on silica monoliths and titanium dioxide. *Environ. Sci. Pollut. Res.* **2016**, *23*, 23742–23749. [CrossRef]
109. Amjadi, S.; Almasi, H.; Ghorbani, M.; Ramazani, S. Preparation and characterization of TiO_2NPs and betanin loaded zein/sodium alginate nanofibers. *Food Packag. Shelf Life* **2020**, *24*, 100504. [CrossRef]
110. Wu, S.; Chen, X.; Yi, M.; Ge, J.; Yin, G.; Li, X. Improving thermal, mechanical, and barrier properties of feather keratin/polyvinyl alcohol/Tris (hydroxymethyl) aminomethane nanocomposite films by incorporating sodium montmorillonite and TiO_2. *Nanomaterials* **2019**, *9*, 298. [CrossRef]
111. Villanueva, M.E.; Puca, M.; Pérez Bravo, J.; Bafico, J.; Campo Dall Orto, V.; Copello, G.J. Dual adsorbent-photocatalytic keratin-TiO_2 nanocomposite for trimethoprim removal from wastewater. *New J. Chem.* **2020**, *44*, 10964–10972. [CrossRef]
112. Siriorn, I.N.A.; Jatuphorn, W. Investigation of morphology and photocatalytic activities of electrospun chicken feather keratin/PLA/TiO_2/clay nanofibers. *E3S Web Conf.* **2020**, *141*, 01003. [CrossRef]
113. Babitha, S.; Korrapati, P.S. TiO_2 immobilized zein microspheres: A biocompatible adsorbent for effective dye decolourisation. *RSC Adv.* **2015**, *5*, 26475–26481. [CrossRef]
114. Johari, N.; Hosseini, H.R.M.; Samadikuchaksaraei, A. Novel fluoridated silk fibroin/TiO_2 nanocomposite scaffolds for bone tissue engineering. *Mater. Sci. Eng. C* **2017**, *82*, 265–276. [CrossRef] [PubMed]
115. Mehrabani, M.G.; Karimian, R.; Rakhshaei, R.; Pakdel, F.; Eslami, H.; Fakhrzadeh, V.; Rahimi, M.; Salehi, R.; Kafil, H.S. Chitin/silk fibroin/TiO_2 bio-nanocomposite as a biocompatible wound dressing bandage with strong antimicrobial activity. *Int. J. Biol. Macromol.* **2018**, *116*, 966–976. [CrossRef]
116. Kazek-Kesik, A.; Peitryga, K.; Basiaga, M.; Blacha-Grzechnik, A.; Dercz, G.; Kalemba-Rec, I.; Pamula, E.; Simka, W. Lactoferrin and collagen type I as components of composite formed. *Surf. Coat. Technol.* **2017**, *328*, 1–12. [CrossRef]

117. Feng, X.; Guo, Y.; Chen, J.; Zhang, J. Nano-TiO_2 induced secondary structural transition of silk fibroin studied by two-dimensional Fourier-transform infrared correlation spectroscopy and Raman spectroscopy. *J. Biomater. Sci.* **2012**, *18*, 1443–1456. [CrossRef]
118. Hu, M.; Lin, D.; Shang, Y.; Hu, Y.; Lu, W.; Huang, X.; Ning, K.; Chen, Y.; Wang, Y. CO2-induced pH reduction increases physiological toxicity of nano-TiO_2 in the mussel *Mytilus coruscus*. *Sci. Rep.* **2017**, *7*, 1–11. [CrossRef]
119. Chen, Z.; Han, S.; Zheng, P.; Zhou, D.; Zhou, S.; Jia, G. Effect of oral exposure to titanium dioxide nanoparticles on lipid metabolism in Sprague-Dawley rats. *Nanoscale* **2020**, *12*, 5973–5986. [CrossRef]
120. Runa, S.; Khanal, D.; Kemp, M.L.; Payne, C.K. TiO_2 nanoparticles alter the expression of peroxiredoxin antioxidant genes. *J. Phys. Chem. C* **2016**, *120*, 20736–20742. [CrossRef]
121. Tucci, P.; Porta, G.; Agostini, M.; Dinsdale, D.; Iavicoli, I.; Cain, K.; Finazzi-Agró, A.; Melino, G.; Willis, A. Metabolic effects of TiO_2 nanoparticles, a common component of sunscreens and cosmetics, on human keratinocytes. *Cell Death Dis.* **2013**, *4*, 1–11. [CrossRef] [PubMed]
122. Gheshlaghi, Z.N.; Riazi, G.H.; Ahmadian, S.; Ghafari, M.; Mahinpour, R. Toxicity and interaction of titanium dioxide nanoparticles with microtubule protein. *Acta Biochim. Biophys. Sin.* **2008**, *40*, 777–782. [CrossRef] [PubMed]
123. Xu, Z.; Grassian, V.H. Bovine serum albumin adsorption on TiO_2 nanoparticle surfaces: Effects of pH and co-adsorption of phosphate on protein-surface interactions and protein structure. *J. Phys. Chem. C* **2017**, *121*, 21763–21771. [CrossRef]
124. Sharif, H.A.; Rasha, A.A.E.; Ramia, Z.A.B. Titanium dioxide content in foodstuffs from the Jordanian market: Spectrophotometric evaluation of TiO_2 nanoparticles. *Int. Food Res. J.* **2015**, *22*, 1024–1029.
125. Wang, Y.Q.; Zhang, H.M.; Wang, R.H. Investigation of the interaction between colloidal TiO_2 and bovine hemoglobin using spectral methods. *Colloids Surfaces B Biointerfaces.* **2008**, *65*, 190–196. [CrossRef] [PubMed]

MDPI
St. Alban-Anlage 66
4052 Basel
Switzerland
Tel. +41 61 683 77 34
Fax +41 61 302 89 18
www.mdpi.com

Coatings Editorial Office
E-mail: coatings@mdpi.com
www.mdpi.com/journal/coatings

www.ingramcontent.com/pod-product-compliance
Lightning Source LLC
LaVergne TN
LVHW070651100526
838202LV00013B/932